INTERNATIONAL RELATIONS AND THE GREAT POWERS

General Editor: **John Gooch**
Professor of International History, University of Leeds

JAPAN

AND THE WORLD SINCE 1868

Michael A. Barnhart
Associate Professor
State University of New York at Stony Brook

Edward Arnold
A member of the Hodder Headline Group
LONDON NEW YORK SYDNEY AUCKLAND

First published in Great Britain 1995 by
Edward Arnold, a division of Hodder Headline PLC,
338 Euston Road, London NW1 3BH

Distributed exclusively in the USA by
St. Martin's Press, Inc.
175 Fifth Avenue, New York, NY 10010

British Library Cataloguing in Publication Data
A catalogue record for this book is available from the British Library

Library of Congress Cataloging-in-Publication Data
A catalog record for this book is available from the Library of Congress

ISBN 0 340 52858 3 (Pb)
ISBN 0 340 52857 5 (Hb)

1 2 3 4 5 95 96 97 98 99

Typeset in 11/12½ Bembo by Scribe Design, Gillingham, Kent
Printed and bound in Great Britain by Mackays of Chatham PLC

To Richard W. Leopold

Contents

List of maps

General editor's preface

Great Powers stand at the centre of modern international relations. The history of their inter-relationships once rested entirely on the formal record of diplomacy, and focussed exclusively on the statesmen, diplomats and ambassadors who shaped and conducted foreign policy. Today, international historians, while still acknowledging the importance of diplomacy, cast their nets more widely. Power rests on many foundations, both tangible and intangible. National wealth is one ingredient of power, but so are ideology and national politics. Personalities have their influence on policies, so that there must always be a place for the individual in the history of international relations; but so, too, do the collective institutional forces of modern state bureaucracies, the military and naval establishments, big business and industry. Together, in varying combinations at different times, these forces and factors find their expression in external policy.

The policies of Great Powers are shaped by their unique amalgams of internal ingredients; but they also have to act and react in a particular but changing international environment. Their choices of policy are affected by the shifting balance between them, and by contingent events: wars and revolutions sometimes offer them opportunities but at other times represent set-backs to a preferred course. The aim of this series is therefore to present the histories of the Great Powers in the twentieth century from these two broad perspectives and to show how individual powers, trying to achieve their own priorities in the regional and global arenas of world politics, have sought to balance the actions and reactions of their rivals against the imperatives of domestic policies.

Michael Barnhart's study of modern Japan emphasises the close relationship between domestic policies and external policy. The structure of domestic politics – the rise and then disappearance of the *genrō* (elder statesmen), the roles of the navy and the army, the emergence of modern political parties – have an important part to play in this story. So too does the world arena on to which Japan erupted in 1904–5. After her emergence from isolation, Japan's is a story in two halves: of regional military opportunism from 1894 to 1945, followed by international economic opportunism under the shelter of the Cold War bipolar system. Japan entered the twentieth century a local military power with great ambitions: as Michael Barnhart shows, she leaves it as one of the economic superpowers whose role in a new post-Cold War international order has to be defined anew.

John Gooch
University of Leeds

1
Introduction

There are few stories of vicissitudes in international affairs to rival Japan's. Nor is there any shortage of images to illustrate those vicissitudes. In 1853, at the moment of American Commodore Matthew G. Perry's entry into Edo Bay, Japan was into its third century of self-imposed isolation from the outside world. Japan had given up the technology of firearms and had not a single steam engine in the country. Fifty years later, Japan was on the brink of a victorious war against Russia that would open the path for its dominion over much of Northeast Asia. It not only had a modern army and navy, but was well along toward manufacturing its own battleships and artillery pieces from its own steel mills. In 1942, Japan's empire spanned almost a fifth of the globe, from northern Manchuria to Burma, nearly to Australia and halfway across the Pacific. Its aircraft were among the most advanced in the world.

By the end of the Second World War, Japan had lost all of its empire. It accepted a new constitution that radically restricted its military capabilities in the future. Its leaders agreed to an alliance with the United States, a rival of nearly a century, that sacrificed a significant degree of independence in Japan's international relations. Yet the postwar years hardly tell a story of defeat. From literally the atomic ashes of the war, Japan rose to become the second greatest economy on the planet, its technological achievements the envy of other industrialized nations and its prosperity of the first rank. In no small part, this economic 'miracle' was aided by Japan's foreign policy choices since 1945, just as those choices before that year led to security, aggrandizement, and eventually disaster.

It is a historian's axiom that nothing is inevitable, but that all things are connected to their pasts. It is especially interesting to examine Japan's foreign relations in this respect. How does one account for Japan's leaders' initial prudence, subsequent folly, and ultimate success?

The appropriate place to begin any answer is in an examination of the leaders themselves. This may seem to be a strange approach at first glance. The Japan of the postwar era had no powerful leaders, no 'great men'.

In fact, several recent studies of Japan openly bemoan the closeted policymaking by consensus that appears to condemn Japan to incremental change, perhaps outright stubbornness to change, in reaction to new international environments. This is nothing new. It is fair to say that of all the powers involved in the Second World War, Japan lacked a single leader of the stature and, more important, authority of any other power, Axis or Allied. Japan had no Roosevelt or Churchill, much less a Stalin or Hitler. Nor can one find a truly dominating individual at any point in modern Japan's foreign relations, including the critical decades of the later nineteenth century.

Nevertheless, the leaders of Japan, and the structure of leadership in Japan, were crucial in making every step, or misstep, in Japan's foreign relations since 1853. For most of those years, moreover, those foreign relations have been more determined by domestic struggles, always for power and often over real policy differences. This study, therefore, focuses upon those struggles when they have affected foreign policy, which is to say since Perry's visit up to the mid-1990s and likely beyond. Put another way, who was in power did make a difference, sometimes a critical difference, in setting Japan's course in the global arena even though (and sometimes because) that person never exercised anything approaching dictatorial powers within Japan.

A second theme implicit in this examination addresses the matter of consensus in Japan's foreign policy making. In the 1970s and 1980s many Westerners wrote of a 'Japan, Inc.', a monolithic amalgam of business, political and bureaucratic leaders who cunningly collaborated to steer Japan to its envious economic performance. In many ways, this image is not too far removed from views during the Pacific War of the Japanese as ants or bees, willingly foregoing individual gain and gratification for the good of the empire as a whole and over the long term. While the earliest years of the twentieth century still permitted a measure of Western admiration for what Japan had so recently accomplished, the appellation 'faceless' still commonly came to foreign minds and tongues.

This preconception does not stand scrutiny. Japan faced a usually hostile and always chaotic international environment from Perry's arrival in 1853 at least until the commencement of the American Occupation in 1945. A national consensus on how to meet such an environment was never sustained, and only very rarely achieved, during these 92 years. The debates over how to meet Perry's challenge resulted in civil war, culminating in the Meiji Restoration of 1868. Within 10 years, the interplay between domestic turmoil and adventurism abroad led to the Satsuma Rebellion. These upheavals, hardly characteristic of a 'hive mentality', are traced in this study's Chapter 2.

The third chapter examines the development of Meiji Japan's early foreign relations within the context of its emerging domestic institutions. The struggles between oligarchs, bureaucrats (civil and military), and legislators were muted but quite real. Often they determined what was possible in Japan's foreign relations, such as who would control Korea, once it was controlled, and what was nearly impossible, such as the delicate peace negotiations ending the Russo-Japanese War. By Chapter four, and the passing of the Meiji Emperor and era, modern Japan faced new foreign challenges in the form of the Chinese and Russian revolutions even as it lost its founders. The result was a new round of internal divisions that saw the emergence of the military, and especially the Imperial Army, as an independent actor in the making of Japan's foreign relations. The Taishō Crisis of 1912 was only the first manifestation of civil–military difficulties. Adventurism in China and serious friction with the United States were likewise outgrowths of internal fractures within Japan's decision-making apparatus during these years. The fourth chapter ends with both army and navy restive under the international settlement for the Pacific negotiated after the First World War, and with the best hope to forge a new, lasting consensus on foreign policy dead by assassination.

The 1920s are the focus for Chapter five. It is, in this study at least, a story of failure. Japan's leaders came to no common conclusion to establish the basis for relations with an increasingly difficult China, much less a hostile Soviet Union and increasingly wary West. They had left behind both the trappings and substance of the old imperial order, abroad and at home. But no substitute had arisen for either during this uncertain decade. By its end, a group of young army officers, with no memory and bare understanding of the accomplishments of the founders of modern Japan, had vowed to complete the destruction of the Meiji system and substitute for it new orders at home and abroad. In some respects, however, their ambitions were the culmination of Japan's attempt to achieve complete self-determination for itself in an uncertain world.

As the sixth chapter makes clear, that world – and Japan's place in it – became even more uncertain in the 1930s. These troubled years saw the sharpest conflicts yet among Japan's leaders, themselves unsure of their country's proper place. Some yearned for the alignments and dispositions of 30 years before, but could never answer satisfactorily how these would fail to compromise Japan's sovereignty in the harsh decade of the Great Depression. Others, particularly within the Imperial Army, strove to make Japan part of a new global order characterized by the rise of regimes totalitarian in all domestic affairs and dedicated to securing dominance of self-sufficient blocs of territory in the global arena.

These conflicts within Japan were not resolved in 1937, with the outbreak of war in China, nor in 1941, when Japanese forces attacked Britain, America, and other Western powers. Army reformers and other, usually civilian, defenders of the older order within Japan fought to a stalemate as Japan's military scored spectacular victories and, swiftly, endured catastrophic defeats. In many ways, though, the surrender of 1945 was a capitulation of the reformers and, as Chapter seven is entitled, a triumph of Japan's conservatives. It was not an automatic triumph. The American authorities of the Occupation years initially intended reforms of their own, different from the Japanese army's scheme of things but no less a threat to those Japanese conservatives. Fortunately for them, the wily Yoshida Shigeru, dominant figure of those years, succeeded in preserving the old order within Japan even as he secured a prudent and profitable place for Japan within the emerging international order of the Cold War, a place firmly entrenched in the aftermath of the Security Treaty Crisis of 1960.

This study's final chapter examines the successes and constraints of Japan's foreign relations in the Cold War era. Here too, however, it seeks to emphasize that those relations were not the automatic or inevitable result of a firmly forged domestic consensus. There have been challenges from the Japanese Left, especially during times of East–West tensions, and more serious attempted revisions from the Japanese Right, especially as Yoshida's way secured for Japan economic triumphs so complete that only the United States had a larger output of goods and services by the 1980s. This story is remarkable, in fact, not that these challenges or attempted revisions existed, but because the international environment for Japan was so tranquil from at least 1960 to the early 1990s. The contrast to every prior period in Japan's foreign relations is stark. With the end of the Cold War order, there are again signs of new divisions and renewed acrimony in the debate over Japan's proper role in the world. It may be that a study of the history of Japan's foreign relations before those halcyon decades will have more contemporary usefulness than is commonly supposed or, perhaps, may be desired.

2
From Perry to empire

Japan's twentieth century began in 1853. The arrival of four American warships under the command of Matthew G. Perry alarmed the Japanese Government and triggered a debate – indeed a revolution – over the nature of that government out of all proportion to their small numbers. As well they should have: Perry's ships represented far more than the young ambitions of a new country with a new frontier on the Pacific Ocean; they were a sign to the Japanese that the West had arrived upon Japan's shores, and that the challenge of the West had to be confronted at long last. The nature of Japan's confrontation with the West would dominate the foreign policy of Japan from 1853 to the present, and so it dominates the story of this book.

In many respects, Japan's response was shaped by its own internal condition. In 1853 that condition was parlous, and as a result the élites of Japan were deeply divided over what their response to the West ought to be. This was not a new story. Every non-Western nation or non-nation has had to seek its own accommodation with the West during the ascendancy of Europe and its daughter, America. What was new was the success that Japan achieved in its ability to find a response that avoided a period of direct Western colonial control. No other major power of Asia managed this feat. How did Japan do it?

The frank answer is that while Japan did avoid becoming a formal colony of the West, it was forced to accept limits on its sovereignty. Those limits outraged Japanese patriots who overthrew the old government in the famous Meiji Restoration of 1868 and established one of their own. These new leaders quickly discovered that the West would not be so easy a foe as the old government in Japan. But they were quick learners and bided their time while enacting domestic reforms that would satisfy the West, which, in turn, would agree to restore Japanese sovereignty. Japan's first search for security in the hostile world of the late nineteenth century was, then, a search for sovereignty.

Perry found a feudal Japan in 1853. Over a hundred *daimyō* – barons – ruled their lands from fortified castle-towns. These *daimyō*, in turn, owed allegiance to the greatest 'baron' of all, the *Shōgun*, who was leader of the Tokugawa family that had united Japan under its control after great battles nearly two and a half centuries earlier.

The Tokugawa had established a rigorous and highly successful system of controls over the *daimyō* in order to maintain power. Part of those controls included the use of the Japanese Emperor. The Emperor was a figurehead without substantial power, but he could trace his family back nearly 2000 years to the first Emperor of Japan, so the emperorship was a highly potent symbol that the Tokugawa used to legitimize their rule. Another part was the Tokugawa policy of *sakoku* – isolation – so that no foreign challenges to Tokugawa supremacy would arise. Except for a tiny research station on a small island off Japan's coast, visited by a Dutch ship once per year to drop off books and instruments, Japan had no contact with the outside world.

For 200 years this system worked. But by the end of the eighteenth century, the Japanese knew of increasing Western encroachments near them. Russian settlements were founded just across the Sea of Japan in Siberia. The Napoleonic Wars of Europe had brought a British frigate to Japan in 1808, where her captain had demanded reprovisioning at gunpoint. But far more ominous than either of these was the coming of the Opium Wars to China beginning in 1839. The West, primarily Great Britain, fought those wars to compel China to open ports to Western traders and to open the country's interior to Western diplomats and missionaries. Western victories yielded these results and more, including an outright British colony carved from Chinese territory: Hong Kong.

The Tokugawa rulers of Japan could imagine no greater nightmare than to have to fight (and lose) their own opium war. When Perry arrived in 1853, therefore, they were hardly caught by surprise. For over 10 years, they had been considering how they could avoid China's fate.

Needless to say, there was more than one possibility. Some advisers to the Tokugawa[1] felt that some kind of an accommodation with the West was inevitable, and that meant reforms within Japan to reach such an accommodation to avoid an opium war against Japan. In the meantime, Japan could build up its military power based on new, Western knowledge. Reform was always risky, though, for those in power. It was time, therefore, to call upon the *daimyō* and ensure that they endorsed the changes necessary while continuing Tokugawa rulership.

Conservatives, however, strongly opposed reforms within Japan to meet Western demands. To be sure, Japan ought to seize Western learning for military uses, but any sign of weakness had to be avoided – China's initial concessions had led to the disastrous opium conflicts. Besides, once it appeared that the Tokugawa were too weak to resist the Western barbarians, why should the *daimyō* want them to retain the title of *Shōgun* (which meant, after all, barbarian-fighting general)? Stalling was a better idea. Unsurprisingly, influential members of the Tokugawa family such as Tokugawa Nariaki, a rising star, held to this view.

The *Shōgun*, Tokugawa Iesada, himself agreed with the conservatives, but he did not believe that Japan could prevail in a war with the West if a no-concession policy brought conflict. His answer to Perry in 1854 offered some concessions, though not many. He opened Shimoda and Hakodate, two remote ports, to the West. These satisfied the American commodore (who may not have understood just how insignificant those ports were), but they horrified Tokugawa Nariaki and the other conservatives. It was clear that the *Shōgun* was going to have a difficult time.

Just how difficult was made clear only 4 years later, when American diplomat Townsend Harris brushed aside the concessions to Perry as virtually meaningless. Harris demanded the opening of major ports such as Edo (now called Tokyo), Osaka, and Kobe. He wanted Western diplomats permanently stationed in Edo, the Tokugawa capital, and rights of extraterritoriality for all Westerners in Japan. Extraterritoriality itself was a major demand. It required that, in all legal matters, Westerners would be subject to their own laws, not Japan's. Finally, Japan would have to agree to fixed tariffs – fixed so as to permit the cheap and easy importing of factory-made Western goods.

The Harris Treaty, in short, demanded much of what China had been forced to yield after the Opium Wars: many of its sovereign rights as a nation. For this very reason, many conservative *daimyō* fiercely objected to the treaty. But the *Shōgun* agreed to it. He still believed that the alternative was to fight and lose a war with the West, to have to yield the treaty's terms anyway, and to probably see the Tokugawa lose power for having lost the war. By the same token, the *Shōgun* realized that his accepting the Harris Treaty would outrage many conservatives, so he sought to deflect criticism from it by having the Emperor endorse the treaty. This clever tactic would put any *daimyō* criticizing the treaty in the unforgivable position of criticizing the holy Emperor.

Indeed, this tactic was far too clever. The Emperor (actually, his advisers) understood the *Shōgun*'s plan and refused to play along. Instead, they began to cultivate ties with leaders such as Tokugawa Nariaki who favoured a much stiffer response to Harris's demands. Their timing was perfect. The old *Shōgun* was dying and, more to the point, childless. Tokugawa Yoshitomi was the closest by blood, but he was only 11 years old. Another contender was the able and vigorous Hitotsubashi Keiki, who just happened to be Tokugawa Nariaki's son. The Emperor's Court elected to take no position on the Harris Treaty and hinted at its favour for Hitotsubashi, a clear victory for the conservative hardliners.

The old *Shōgun*'s advisers could not let this pass. They quickly confirmed young Yoshitomi as *Shōgun*, approved the Harris Treaty, and arrested Tokugawa Nariaki and Hitotsubashi. This power play failed terribly, touching off a firestorm of protest from the conservative *daimyō* and igniting demonstrations from Japanese patriots, especially among the *samurai*, Japan's warrior class. Put another way, the play began a series of events that led to the Meiji Restoration.

The men who would lead that Restoration were *samurai*, warriors in the service of the *daimyō* who had taken part in no wars for over 200 years. Some, such as Saigō Takamori, kept close to their tradition and began to consider direct violence, against the Westerners (who were now admitted into Japan under the Harris Treaty), to be sure, but also those Japanese who had permitted such an outrage. Others, such as Ōkubo Toshimichi, Itō Hirobumi, and Yamagata Aritomo, had mastered the arts of government and administration, not just the sword. They were as patriotic as Saigō, just more patient, preferring to win over the cooperation (and resources) of their *daimyō,* especially those in their own Satsuma and Chōshū provinces, rich lands in the south which had lost the wars against the Tokugawa two centuries earlier and which still distrusted the Tokugawa family. All of these *samurai* patriots, however, agreed that the Tokugawa shogunate had violated the sacred land and spirit of Japan by truckling to the foreigners and, in so doing, had disgraced the Emperor. Their rallying cry thus became *sonnō jōi* – revere the Emperor; expel the barbarians – and foreign policy considerations thus became a key focal point of the Meiji Restoration.

While hotheads like Saigō plotted, Ōkubo, Itō and Yamagata moved to recruit the support of the *daimyō*. They were hugely successful. In 1862, Satsuma sent troops to Kyoto, the ancient capital where the Emperor still resided. Satsuma planned to use the Emperor to put pressure on the *Shōgun* to adopt a harder line toward the foreigners (Satsuma *samurai* would murder a British diplomat later that year as Saigō's ideas took hold among the more radical *samurai*) and a faster pace for domestic reforms, including a military build-up. Chōshū was even more direct, closing the Straits of Shimonoseki in 1863 and firing on Western ships passing through. Local Western (usually British) forces attacked Satsuma and Chōshū in retaliation as a large Western fleet was gathered for widespread operations against Chōshū in 1864.

The dying *Shōgun*'s advisers faced a choice: civil conflict with these provincial patriots, especially in Chōshū, in order to suppress their antiforeign activities, or a general war with the West. The choice seemed easy, particularly since the West could be relied upon to punish Chōshū anyway. The strategy worked quite well. The West punished only Chōshū in a strong display of Western firepower. The other *daimyō*, impressed, moved to support the *Shōgun*. *Samurai* who had been attacking Westerners were rounded up and imprisoned or worse. Even Satsuma used force to disband a group of radical *samurai* who had appointed themselves the Emperor's guard. The surviving radicals fled to Chōshū. But they found scant refuge. The *Shōgun* proposed to invade Chōshū to crush the last challengers to its policy of accommodation. To avert that invasion, the *daimyō* of Chōshū ordered the radicals to commit suicide. This gesture spared Chōshū from the *Shōgun*'s forces for the moment, but triggered a bitter, complex struggle for power within Chōshū that led to the ascendancy of those patient patriots, such as Itō and Yamagata, who would be pivotal in the coming Meiji Restoration. In any event, the hotheads of *jōi* had been put into the background and the *Shōgun's* advisers were in control.

At this point they miscalculated. Having assumed by early 1866 that the Western threat was in abeyance, the *Shōgun* resolved to deal with the Chōshū problem once and for all: an invasion to crush that province. It was an ill-timed and uninformed decision. The British, with a better understanding of the uses of symbolism in Japan, had just called for the Emperor publicly to approve the Harris Treaty and others like it negotiated with the West. The Tokugawa, seeing the implicit challenge to their authority, compelled the Emperor to so approve, a humiliation not lost upon either the Emperor's Court nor upon the *daimyō*, some of whom began to think that resistance to the West might be the right and honourable course after all. As well, hardline reformers such as Satsuma newly understood that the Emperor's Court could not be used to pry needed reforms from the *Shōgun*. More direct action would be necessary to preserve Japan (and the Emperor) in the face of a Western threat that Satsuma saw as more real than ever. Moreover, Satsuma was appalled at the idea that the Tokugawa would risk civil war by attacking Chōshū while that foreign threat was so great. After Chōshū turned back the Tokugawa forces, Satsuma's and two other provinces' *daimyō* allied with Chōshū to fight the Tokugawa – Japan's civil war had begun.

It was over almost at once; by the autumn of 1868 victorious provincial leaders proclaimed the overthrow of the Tokugawa *Shōgun* and declared Emperor Meiji restored to his rightful powers. Of course, the Emperor was still very much the figurehead, still with the power of legitimacy but little direct control over policy, foreign or otherwise. This was probably just as well, for the new Meiji oligarchs, most of them *samurai* from Chōshū and Satsuma, had far-ranging plans to scrap Japan's feudal order, for one overarching goal: to rid Japan of the 'unequal treaties' such as the Harris Treaty and be able to confront the West as a sovereign equal. The search for security had begun in earnest.

The new Meiji leaders were determined to remove the West's restrictions on Japanese sovereignty. To do that, they realized that they would have to satisfy Western standards. Ironically and conveniently, these standards pointed the way to how Japan ought to be reformed internally to consolidate Meiji rule. As a result, the new Meiji leadership put domestic reform first. Once that reform was accomplished, Japan's full sovereignty could be recaptured. In many respects, therefore, domestic and foreign policies were inseparable. As well, modern Japan's foreign policies were begun, and have largely continued, as adjustments to external conditions, from the nature of international systems to the actual rules of diplomacy, both customarily defined by the West.

Japan's domestic reform was achieved by the Meiji oligarchs' intensive study of things foreign: not just Western technology, but also European and American institutions, especially governmental institutions from armies to schools. Indeed, the Meiji leaders thought such study so vital that they took it upon themselves personally. In late 1871 the Iwakura Mission, made up of such great figures as Ōkubo, Itō, Kido Kōichi, and Iwakura Tomomi, an official of the Emperor's Court, departed on a multi-year study of the West.

Two foreign-policy crises forced their early return to Japan. Yamagata, a military genius from Chōshū who had been instrumental in defeating the Tokugawa, was eager to create a new Imperial Japanese Army as rapidly as possible. To that end, he had instituted the Western practice of conscripting able-bodied males on a mass basis. Logical enough on its face, conscription had two potentially explosive consequences in early Meiji Japan. Peasant families, acutely in need of strong helpers for a labour-intensive style of agriculture, were distressed at losing their young men. Even worse, though, was *samurai* mortification at losing their monopoly over matters military. The *samurai* rightly saw conscription as the death knell for their status as a class. Many of them moved toward resistance, gathering under traditionalist Saigō in Satsuma.

Saigō and the *samurai* believed they had found the occasion to prove their value in the Korean crisis of the early 1870s. The Korean situation was, to understate, both curious and complex. By the 1870s, the Koreans were coming under Western pressure to open trade with the outside world and to reform their government. As with all other nations touched by the West, Korea had entered a time of turmoil. Koreans themselves disagreed over how best to meet the Western challenge. Some thought that Japan's response – an internal revolution in order to carry out thorough reforms at home – was the best way. Other Koreans still looked to China, which had exercised a form of suzerainty over Korea for centuries, and elected to cling to the old ways. These conservative Korean leaders had succeeded in blocking Korea's recognition of Japan's new Imperial Government by claiming that only the Chinese Emperor deserved that title. This insult was not to be borne, argued Saigō. Insolence from the West might have to be swallowed for the time being. But Korea was another matter.

It was not another matter at all, responded cooler heads such as Ōkubo, Itō, and Yamagata. A Korean expedition would invite Western moves against Japan, and at a time when the Meiji regime was still in its infancy and the Imperial Army unready. Ōkubo's arguments carried the day, but at the cost of driving several oligarchs, including Saigō, to resign from their positions in the Meiji regime. In Satsuma, Saigō led his unrepentant *samurai* in the Satsuma Rebellion of 1877, one that Yamagata's Imperial Army would crush in short order. Once again, Japan had elected to placate the West, even at the cost of another, if small and brief, civil war. It would cost Meiji Japan talent, too: on 14 May 1878 embittered followers of the dead Saigō murdered Ōkubo.

On the other hand, the (surviving) leaders of Meiji Japan would never again face a significant challenge to their position from other Japanese. The West was another matter.

The Meiji oligarchs clearly understood that the departed *Shōgun*'s inability to protect Japanese sovereignty – his humiliation in having to accept the unequal treaties – had been one key to the downfall of the Tokugawa. They were, therefore, determined to secure treaty revision as quickly as possible. At the same time, these patient patriots were under no illusions that quick revision was possible. Moreover, some changes wrought by the treaties, such as the opening of Japan to the world, the Meiji leadership positively welcomed.

But the institution of extraterritoriality and the imposition of tariff controls – these were to be suffered only so long as necessary. It is ironic that the Meiji leaders' attempts to shorten the waiting led to renewed agitation against them inside Japan.

These leaders, many of whom had been on the Iwakura Mission, knew that the West would not consent to revision until Japan had met Western standards, especially in the field of law. Indeed, Western rebuffs to Japanese attempts to renegotiate the unequal treaties in 1878 and 1880 reinforced this argument of putting domestic reform first. Japan needed a 'modern' set of laws, complete with Western-style training for lawyers and judges. At the pinnacle of these efforts would stand the Meiji Constitution (which, among its other features, would absolutely reserve the conduct of foreign policy to the Emperor – that is, to the Emperor's Cabinet). Hardline conservatives within Japan bridled at these changes. Discontented reformers grumbled over the degree of highly centralized authority in the new, constitutional government. Unfortunately for the Meiji oligarchs, their constitution had to include provisions for a (weak) legislature, the Diet. In this way, the Meiji leadership faced a paradox. To secure treaty revision required Western-style laws and a Western-style constitution, hence a Western-style legislature. But the birth of that legislature would inevitably lead to intense pressure for immediate treaty revision and great embarrassment for the Meiji leadership if revision had not been achieved by the opening of the Diet. The reason was simple. One of the leaders in the Diet, scheduled to first meet in 1890 (the year after the constitution would be promulgated by the Emperor) was Itagaki Taisuke. Itagaki, formerly an oligarch himself, had broken from the Meiji leadership in 1873, ostensibly because he had supported Saigō in favouring an expedition to Korea. Another, perhaps more sincere reason for his bolt was his place of birth: Tosa province. Itagaki, with some justification, complained that all the top positions of influence in the Meiji regime had gone to Chōshū and Satsuma men. In any event, Itagaki and his so-called Liberal Party (*Jiyūtō*) were determined to attack the Meiji leaders for their failure to secure treaty revision as soon as the Diet was convened. A second political party, the *Kaishintō* (Reform Party) was led by another dissident leader, Ōkuma Shigenobu from Saga, and entertained similar motives. To escape their paradox, the Meiji leaders attempted to secure limited, provisional revision of the unequal treaties before 1890.

This strategy backfired. The attempt had been in the hands of oligarch Inoue Kaoru and his protégé, Foreign Minister Mutsu Munemitsu from May 1886 to April 1887. Inoue, a firm believer in the need for Japan to aggressively adopt foreign ways, if necessary from foreign hands, had judged acceptable a revision of the treaties that would have both Japanese and Western judges sitting at cases involving Westerners until such time as Japan had completed creating a legal system (with trained personnel) that the West could have confidence in. This stance may have succeeded diplomatically, but inside Japan it was political dynamite: a tangible, visible symbol of Meiji Japan's continued lack of full sovereignty.

Prime Minister Itō, who had emerged as the leader of the oligarchs, dealt with the outcry with characteristic direction and deftness. On the one hand, he had the Emperor issue laws authorizing suppression of dissent (a move entirely legal, since the new constitution would be promulgated only by 1889). On the other, he named Ōkuma of the *Kaishintō* his new Foreign Minister, and so responsible for handling the hot issue of treaty revision.

A chagrined Ōkuma found the West just as obdurate and pressure inside Japan for immediate change just as insistent. His attempted remedy was to reach essentially the same compromise Mutsu had: foreign judges would still sit in Japanese courts in cases involving foreigners. But Ōkuma tried to finesse this obvious difficulty by keeping his negotiations entirely secret. How he intended to keep the new treaty terms hidden after those negotiations ended remains a mystery, because in July 1889 the London

Times leaked word of those terms, causing an instant uproar in Japan. A flurry of cabinet meetings followed, climaxed by an Imperial Conference (so-called because it was held in the presence of the Emperor) in October 1889. The new Diet's first meeting was only a year away, a fact Ōkuma used to defend his plan: it was his revisions or nothing by the time the Diet began. But most of the Meiji leadership preferred nothing, compelling Ōkuma to break off talks with the West. This move proved too late for one radical proponent of treaty reform, who severely injured Ōkuma with a bomb.

The early sessions of Japan's Diet, beginning in November 1890, surprised no one in demanding instant and complete revision of the treaties. But the Diet did disappoint the Emperor's Cabinet by delaying the institution of new civil and commercial codes – exactly the ones needed in place if the West were to agree to revision. Part of the reason for the Diet's contrariness was pride, to be sure, but a good deal of spite against the Meiji oligarchs, whose cabinets refused to allow (Diet party) politicians admission, was at work as well.

In this unpromising environment, Mutsu was brought back as Foreign Minister to make a second try in 1893. Xenophobic demonstrations swirled around (and sometimes in) the Diet. But Mutsu's new attempt paid off. In July 1894, Great Britain agreed to relinquish the right of extraterritoriality in 5 years. There were two keys to Mutsu's success. By the spring of 1894, Japan and China were at sword's point over Korea. The impending crisis drew the attention of the West, especially a Great Britain increasingly anxious over the spread of Russian influence in Northeast Asia and over the prospect that the Sino-Japanese showdown would benefit Russia in the end. The British decided to begin cultivating Japanese goodwill as a result.

The second factor aiding Mutsu was the delayed but at last imminent completion of Japan's new legal system. The critical element of that new legal system was its personnel. Japan had to have Japanese trained in law and justice to the West's (and, for that matter, to the Diet's) satisfaction. Before 1894, this had been a difficult task. The Meiji leaders were hardly satisfied with sending promising Japanese legal minds overseas to study. But the only real alternative was the creation of a Japanese system of higher education that would meet the standards of the West. Fortunately, Japan was blessed with a society that highly valued formal education. Unfortunately, each Meiji leader correctly saw that whoever controlled the universities that educated the future lawyers and other élites of the bureaucracy of Meiji Japan effectively would control much of the government itself. For this reason, Ōkuma had founded Waseda University in 1882, after his exile from the ranks of the Meiji oligarchs.

Itō undertook the logical counter to such a challenge. He ensured that graduates of Tokyo Imperial University had the best chance to secure top-level (that is, policy-making) positions in the government by adopting a rigorous examination system that permitted barely one in ten candidates to pass. This system, moreover, was applied to every government ministry in the Meiji regime, including the Foreign Ministry, which had its own, specialized examination. As a result, the members of Japan's bureaucracy were a close-knit, even insular club, many of whom knew each other from their school days.

In any event, the invention of the higher education and examination systems had led to success in overthrowing the unequal treaties. But Foreign Minister Mutsu barely had the opportunity to relish his accomplishment before Japan found itself at war with China. The cause was a familiar one: difficulties in Korea. For over a decade, Korea had wrestled with the problem of how to address the West's challenge. Most members of the Korean Court preferred a gradual (some might say snail-paced) accommodation so as to minimize domestic turmoil. But there was a visible, active reform element, led by Kim Ok-kiun. Unsurprisingly, these Korean reformers were drawn to Japan's way of dealing with the Europeans, swift and thorough reform including the wholesale change of domestic government. In fact, these reformers often were drawn to Japan itself, studying there and gaining support from private Japanese (such as students of the famous Fukuzawa Yukichi) who believed in their cause. Predictably, the Korean Court increasingly sought support from conservative China.

This situation was a recipe for trouble, which came in the form of the Seoul Uprising of December 1884. Led by reformist Koreans, the uprising nevertheless had clear Japanese support, from Japan's Minister to Korea, Takezoe Shinichirō, reaching as high as Inoue Kaoru. These connections turned out to be necessary, because the Korean plotters failed miserably in their attempt to seize power themselves. With Japanese military assistance, however, they were able to seize the Korean King and kill many of his ministers. Unhappily for both reformers and Japanese, these actions spurred Chinese military intervention under the energetic (and ambitious) Yuan Shih-k'ai. When Yuan's troops entered Seoul to restore order, crushing the reformers' attempted coup and compelling the Japanese legation to flee the country, the Japanese Cabinet under Prime Minister Itō found itself in a crisis.

Itō's reaction was characteristic. However much Japanese reformers might wish to see their ideas, the ideals of the Meiji Restoration, spread throughout Asia, they could not be permitted to jeopardize the success of that restoration inside Japan. Itō went to China, where he negotiated a convention pulling both Japanese and Chinese troops out of Korea. Ostensibly, this mutual withdrawal was very much to Japan's tune.

Actually, the Li-Itō convention was a Japanese surrender. China's claim to suzerainty over Korea was not contested, nor was China's long-term posting of the aggressive Yuan to Korea as adviser to the Korean Court. Inside Japan, Itō oversaw a crackdown on Korean activists and their Japanese sympathizers. Kim, for example, was ordered into exile (though he would linger on in Hokkaido until 1889).

Itō's wider audience for these moves was the West. He, and most other Meiji oligarchs, had no desire to roil Asian waters while Japan's treaty revision was in progress. Quite the contrary, he wished to demonstrate that Japan was capable of Western-style diplomacy and, unlike the *Shōgun*, of maintaining domestic order in the interests of that diplomacy. Why, then, war with China over Korea at just the moment treaty revision was about to be secured?

The simplest answer is still a complex one. Three factors combined in the early 1890s to lead to Japan's first foreign war and territorial expansion onto the Asian mainland. To begin with, China's highly aggressive stance in Korea was quite provocative. Alas for cautious leaders such as Itō, China's provocation came at exactly the time that Japan's final domestic settlement, the Meiji Constitution, saw the convening of the Diet and, hence, the creation of a highly visible forum for critics of the government. These critics wasted no time in attacking the Meiji oligarchs for their submissiveness over Korea. Finally, a new, Russian presence arrived in Northeast Asia with the impending completion of the Trans-Siberian Railway just north of Korea.

Any spark might have ignited the Korean tinderbox. As it turned out, there were two. In March 1894, Kim Ok-kiun was murdered in Shanghai. Chinese authorities allowed his body to be returned to Korea – not a gesture of mercy, as the Korean authorities promptly dismembered it as a warning to other reformers. This stratagem only half worked. Korean reformers did not take to the streets (though Kim's fate aroused tremendous protests among sympathizers in Japan), but Korean traditionalists – who believed that the Emperor's Court had actually gone too far in the direction of change – rose in what became known as the Tonghak Rebellion. The Korean monarchy quickly received Chinese forces to put down the insurrection, a clear violation of China's 1885 agreement with Japan.

Even then, it was possible that leaders such as Itō and Mutsu might have avoided war. But Yamagata was convinced that Japan had to be assertive. Yamagata increasingly saw Russia, not China, as the real threat to Japan's security. Indeed, he believed that China was fated to soon be the subject of Western rivalries, not an independent actor at all. If Japan were to fend off those rivalries, it was best to participate in them. What surer way to finally mark the end of the unequal treaty era? As to the point, Yamagata was confident that his Imperial Army was prepared for a test of arms with China. And his cronies in the Imperial Army were in a position to emphasize the size of Chinese forces entering Korea and downplay the possibility of a diplomatic settlement there. That many Diet members were clamouring for intervention only strengthened Yamagata, who ironically detested the legislature and sought to minimize its (budgetary) influence over the growth of the army. On 1 August 1894, Japan declared war on China.

This date serves well as the inaugural of Japan's second search for security through a search for empire. Yamagata's logic would be at the centre of Japan's foreign relations from 1894 to 1945. Throughout those 51 years, Japan would attempt to defend itself through the exercise of imperial power beyond its simple national borders. It would do so through the calculus of empires, considering weaker neighbours subjects for control, lest other strong empires control them instead and use their locations and resources against Japan.

To the amazement of every nation except Japan, the Sino-Japanese War was a one-sided affair. Japanese forces ousted China from Korea within weeks. By October 1894, the Imperial Army had entered Manchuria and soon seized strategic Port Arthur on the Yellow Sea. Yamagata favoured continuing the assault into China proper (Manchuria still being an incompletely assimilated region of the Chinese Empire), but Itō would have none of it.

Itō's characteristic caution stemmed from two sources. Domestically, he was worried about the colossal cost of the war effort. The war's immediate expenses were over 160 million yen, a sum that was double that of total governmental spending for any peacetime year. Japan was in a position to meet this extraordinary need in large part due to a series of wide-ranging reforms in taxation and banking engineered by Finance Minister Matsukata Masayoshi and the overwhelming popularity of the war with the Japanese public, which subscribed so generously to war bond issues that the government did not have to raise taxes, nor seek to borrow from overseas. Even so, Itō and Japan's financial leaders understood that the war was bringing inflation, with its hidden burdens to the people, and tax hikes in the future, to retire the government's war bonds as they came due.

Internationally, Itō had anticipated the rising concerns of the West. Indeed, he had assured the European powers that Japan would not extend hostilities into central and southern China, such as Shanghai, under any circumstances. A Japanese force did occupy the port city of Weihaiwei on the Shantung peninsula, however, a move that forced China to sue for peace terms and the Western powers to hint that it was time for the fighting to stop.

There was relatively little disagreement in Tokyo over what those

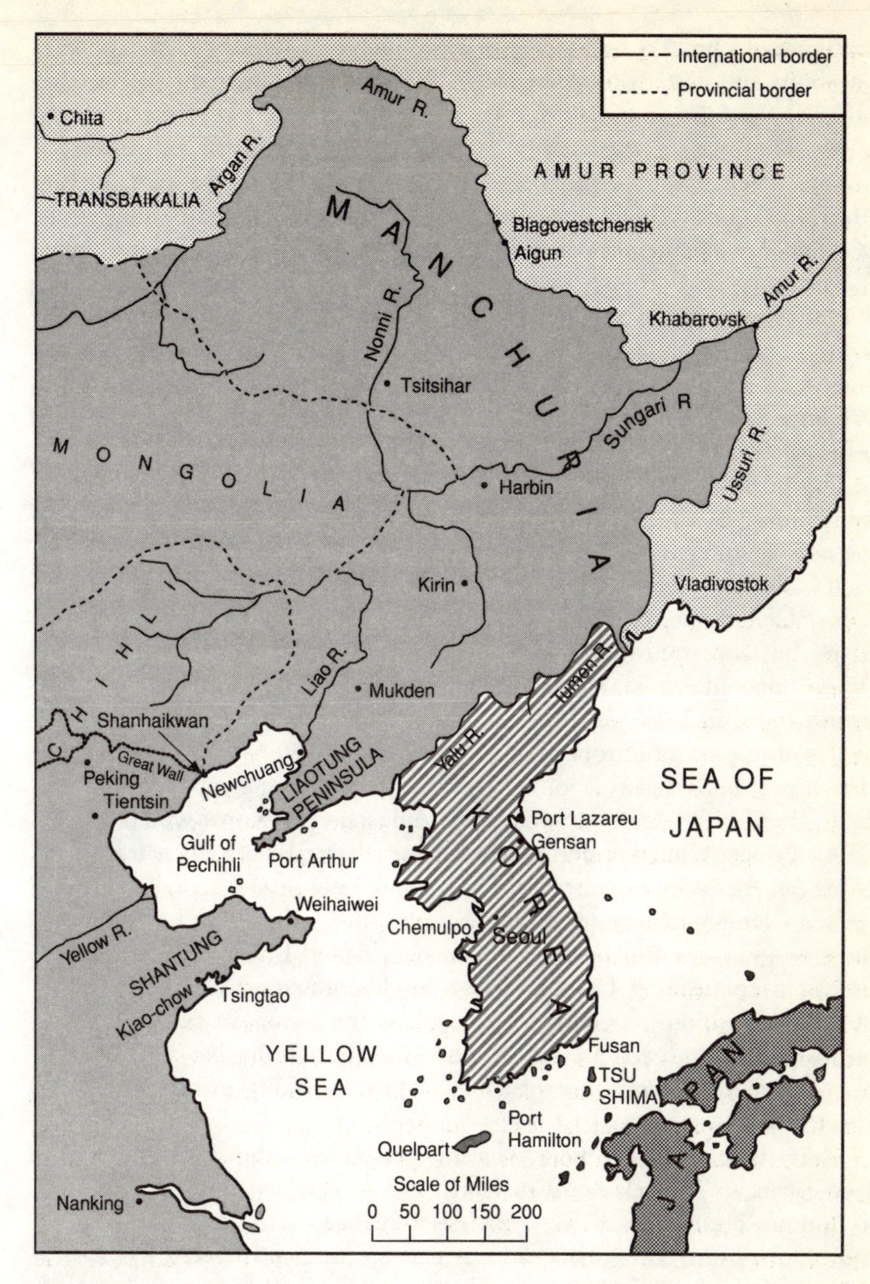

Map 1 Northeast Asia 1894–5

terms should be. To be sure, everyone, including Yamagata, feared the possibility of Western interference in the final peace settlement. But the prevailing sentiment argued that Japan should attempt to extract as much as possible from the thoroughly defeated Chinese, if for no other reason than to determine how much the West would allow to be kept. In the April 1895 Treaty of Shimonoseki, Japan obtained Chinese recognition of Korea's independence (opening the way for Japanese domination), plus the territory of Formosa, the Pescadore Islands, and the key Liaotung peninsula jutting into the Yellow Sea, in addition to a cash indemnity of slighty over 80 million yen. Tellingly, the treaty also included 'most-favoured-nation' treatment by China for Japan: Japan now had joined the Western treaty system as actor, not subject. Chinese sovereignty would be sacrificed, in some measure, for imperial Japan's gain.

The question now was the West's reaction. That reaction was determined more by the global nature of Western imperial rivalry at the end of the nineteenth century than by the peculiarities of the situation in East Asia itself. Globally, the central fact of the 1890s was Russia's seemingly inexorable expansion not only eastward, into China and Korea and towards Japan, but also southwards, into the Persian Gulf. In both instances, the British opposed Russia, having a keen interest in China and the Gulf (and British India, in between). At first glance, it might have appeared possible for Japan to pit one European rival against the other. Actually, neither the British nor the Russians favoured Japan's successes in the war. On the one hand, the British feared that a Japanese annexation of Korea would trigger Russian military intervention that would accelerate Russia's penetration of Northeast Asia. On the other hand, if Japan did not seize Korea, but merely ejected Chinese influence from that country, the path to Russian influence there was likewise laid open. More fundamentally, Britain was concerned that an overwhelming Chinese defeat, by demonstrating China's helplessness, would expose all China to partition as the European powers scrambled for influence, even pre-eminence in the Far East. In that case, Britain, with commercial dominance of the rich Yangtze Valley in central China, stood to lose the most.

More straightforward were Russia's concerns. In the years before the Sino-Japanese War, elements in Korea had turned increasingly to Russia to counter both Chinese and Japanese influence there. It was an opportunity for power that Russia did not pass by, one endangered by Japan's decision for war and subsequent victories. As well, Russia, too, opposed a swift partition of China, confident that Russian influence over the Chinese Government was growing steadily. Why settle for a piece of China in any partition when the whole prize appeared to be within reach?

Nevertheless, Russia was far more inclined to act against the terms of

Shimonoseki than Britain. By the spring of 1895, the British public, and government too, was coming to see Japan as a useful counterweight to Russian influence in East Asia. Besides, the Japanese had scrupulously kept the fighting away from central China, just as they had promised. The Russians, however, could not tolerate a Japanese territorial foothold in northern China, where Russian interest (and influence) was greatest. There was some debate in St Petersburg over whether it might not be wiser to accept the partition of China, which would be begun by Shimonoseki's ceding Liaotung to Japan, and, in fact, join Japan in carving up China. But Finance Minister Sergei Witte, a firm believer in Russia's bringing all of China to heel, successfully insisted upon resistance to any Japanese presence in any Chinese territory, even if that meant Russian naval action against Japan.

This was an exceptionally strong stand, one that Britain could not adhere to, even though the British themselves were unhappy at the prospect of Japan acquiring Chinese territory. But France, drawn increasingly close to Russia in Europe because of rising German power, associated itself with the Russian protest against Shimonoseki. And Germany, angry at Japan for insisting upon terms that might have driven the Russians and British together, gladly joined in what became known as the Triple Intervention. The three powers' protest reached Japan on 23 April 1895, barely a week after the Treaty of Shimonoseki had been signed. Japan would have to give up control of the Liaotung peninsula.

Foreign Minister Mutsu, along with Yamagata and the other leaders of Japan's army and navy, understood perfectly that their country was no match for Russia itself, let alone a Russia joined with France and Germany. They attempted to salvage the occupation of only Port Arthur, yielding the rest of Liaotung. But the three powers would not budge. All Liaotung was returned to China.

Even so, Japan had won a resounding victory on the field of international relations. The West came to regard Japan as one of the powers, though most Western observers were not yet willing to grant Japan great power status. Japan had removed Chinese influence from Korea thoroughly. Yamagata was able to trade upon its recognition as a power and removal of that influence to secure an accord with Russia in the spring of 1896 allowing both nations to station troops in Korea.

But the Triple Intervention provoked a political firestorm within Japan. The Japanese people, after all, had read account after account of their forces' magnificent victories and steady progress. They had contributed substantially to the war's funding; even the poorest Japanese had tapped their meager savings to contribute to the glorious enterprise. To be humiliated by the West – that old theme – was especially bitter after such euphoria. Needless to say, members of the Diet hostile to the Meiji oligarchs, especially Yamagata, were only too pleased to ride this wave of public outrage.

Ordinarily, Yamagata and the other Meiji leaders could have ignored the impassioned speeches in the Diet. Technically, the Diet held little power, even over the budget, under the Meiji Constitution. If it refused to pass a new spending bill, the old one was continued, at the same spending levels, into the next year. But in the aftermath of the war, new spending, especially on the military, appeared to be in order. Indeed, spending on the new prizes of empire, Korea and Formosa, for their economic development, was imperative and substantial. That meant dealing with the Diet because only the legislature could approve spending increases beyond what the old budget allowed. However, Itō and Yamagata had two very different ideas of what sort of deals to make in order to gain these spending increases. Their split over this issue would prove even more decisive for the history of Japan's foreign relations in the early twentieth century than the Triple Intervention that provoked it.

Note

1 One example was Abe Masuhiro.

3
Player in The Great Game

The two decades from the close of the Sino-Japanese War to the eve of the First World War marked the heyday of what historians have called the Age of Imperialism or what a contemporary just as accurately labelled 'The Great Game'. To be a power was to have an empire, and the European powers clashed repeatedly outside Europe as they defined the scope of their empires. Increasingly, by the 1890s, the arena of those clashes was Asia, particularly China. Indeed, it was Japan's victory over China that paved the way for an intense round of competition for 'slices' of the Chinese 'melon' during these years. And, at the end of this period, it was Japan's determination to use the Chinese Revolution to secure great influence over all of China – to become the dominant imperial power in East Asia – that helped lead to an American challenge to the diplomacy of imperialism and to a series of American confrontations with Japan.

As well, these two decades saw the maturation of the Meiji state within Japan. The oligarchs so crucial to the Meiji Restoration had hardly faded from view. In fact, they were by this time seasoned and successful statesmen, a fact formalized in the term used to label them, the *genrō*. But after the promulgation of the Meiji Constitution, they no longer had sole possession of the titles or instruments of power. The Diet was one rival, where Itagaki Taisuke had organized the *Jiyūtō,* and Ōkuma Shigenobu the *Shimpotō* parties, respectively. Another, often overlooked, power centre was the growing Meiji bureaucracy in the civilian and military ministries. Here, too, there was change over these nearly 20 years.

The *genrō* were the glue that held the Meiji state together. The Diet could veto budgets, but it could not propose them, and it had very little power over any other aspect of governing. As a result, the political parties devoted most of their energies to making deals with the *genrō,* who controlled the executive and administrative powers. Likewise, the bureaucrats in their various ministries, such as the Finance Ministry or the Foreign Ministry, had the expertise to run the country. But they were trained at government-run universities, such as Tokyo Imperial University, usually under the sponsorship of a *genrō*, and so the civilian bureaucrats were hardly of a mind to challenge Meiji leadership. The potentially most threatening 'bureaucrats', of course, were the military – the army and navy ministries. But here, too, the service academies were operated by the Meiji Government and the *genrō*, especially Yamagata, had close ties with the senior officers. While the *genrō* were in charge (which is to say, while they were alive), the Meiji state was completely stable.

So was Meiji society. The years from 1895 to 1915 were remarkable through most of the West for their extraordinary turbulence as Europe and America adapted to the Industrial Revolution. Labourers often went on strikes against their employers, and often these strikes turned violent. Governments wrestled with the problems of how to regulate the emergence and operation of gigantic corporations. Sometimes, governments wrestled with their own citizens as political crimes and even assassinations rose in frequency and fever.

Japan was spared much of this turmoil in these years, primarily because of the mediating influence, conservatism, and wisdom, of the *genrō*. Many former *samurai* had been absorbed into the administrative apparatus of the state. Landholders, that is, the rural élite, were entitled to the vote,[1] ensuring that the Diet would look after their interests. The rising group of industrialists either worked directly for the government (especially in military-related industries such as steel) or benefitted handsomely from the government's assistance to bring Japan into the modern world economically. The lot of the common person in Meiji Japan was hard, it is true, whether one remained on the farm or moved, as many Japanese did during these years, to the growing cities. But the fact remains that the cities were growing because Japan was modernizing. Factory work was tedious, often exhausting, but in relative terms it paid quite well. There were some demonstrations, by both peasants and workers, from time to time, but when these failed to die out of their own accord, the Home Ministry's police were quick to put an end to them. For the vast majority of Japanese, then, political issues (and their ability to influence those issues) were as remote as the moon and concerns about foreign relations barely existed. This condition, too, contributed to the stability of Meiji Japan.

But the *genrō* were mortal. By 1912 some had passed on, and those remaining, such as Yamagata Aritomo, were more and more passed by in the policy-making process, another trend that would gather momentum, with fateful consequences, in the years to follow. As a society, Japan would remain remarkably stable throughout the new twentieth century. Within Japan's élite, however, the end of the *genrō* meant the end of consensus on Japan's relations with the rest of the world.

The first signs of dissent among the *genrō* came as early as the aftermath of the Triple Intervention of 1895 and that dissent opened the way for other political leaders to gain influence over the making of Japanese foreign policy. Both Diet parties saw a heaven-sent opportunity to gain influence by attacking the government's submission to France, Germany and Russia. Although the Diet did not have the power to ratify (or refuse to ratify) the actual peace treaty, it did have the ability to block the government's requests for increased military spending.

Alarmed at the prospect of domestic discord in the face of the Triple Intervention, Prime Minister Itō, one of the leading *genrō*, moved to create harmony by allying himself with the *Jiyūtō* (Liberal Party) by making its leader, Itagaki, Home Minister. This was a revolutionary step, the first time that any of the *genrō* had agreed to share power with a party leader. For its immediate purposes, Itō's move worked, but at high price. His budget bills passed the Diet. These included substantial military expansion – a near-doubling of the size of the Imperial Army and a 10-year naval programme that would build a fleet of six battleships and supporting vessels, larger than any other (European) power would have in Asian waters.

But Itō's alliance alienated nearly every other political leader in Japan. Matsukata Masayoshi, a *genrō* himself, decided that two could play Itō's game; he allied with Ōkuma's *Shimpotō* and demanded seats in the cabinet for himself and his new associates. Yamagata was appalled that Itō had opened the way for any political parties to share in real power, and was particularly angry that Itō had given party-man Itagaki the Home Ministry, where Yamagata had built up that ministry's bureaucracy into a powerful, conservative, and (so far) autonomous force. In reaction, Yamagata moved to prevent Itō and Itagaki from removing top bureaucrats in the Home Ministry, and in other ministries – including the military. Here, Yamagata instituted an extra safeguard for bureaucratic autonomy: no serving army or navy minister could belong to any political party. Yamagata, in fact, instituted a wide-ranging set of controls to isolate the ministries – the bureaucrats – from the politicians and to remain bedrocks of conservatism.

These discordant notes ensured that Itō would not remain Prime Minister for long. Ever the cautious conciliator, Itō proposed to bring Matsukata and Ōkuma into his cabinet as well. But Itagaki would not serve alongside Ōkuma, leader of the rival party in the Diet. Faced with the prospect of chronic strife with a fellow *genrō* and Ōkuma's *Shimpotō*, Itō chose to resign. This step gave a turn to another *genrō*–party leader combination: Matsukata, who became Prime Minister, and Ōkuma, who was named the new Foreign Minister. The Matsukata Cabinet had a significant foreign policy success, putting Japan on the gold standard, a move that accomplished symbolic equality with the West in the fiscal realm. But otherwise the cabinet was not well regarded. The other *genrō* were not comfortable with Ōkuma as Foreign Minister. As a party leader in the Diet, Ōkuma had cultivated voter support by casting himself as a patriot and expansionist. He had often called publicly for a strong stand against the West, in marked contrast to the cautious policy that the *genrō* preferred. Itō was sufficiently concerned in 1897 to warn Ōkuma not to protest the United States' imminent annexation of Hawaii, which had a significant number of Japanese immigrants. As it turned out, Matsukata and Ōkuma were not long in office, victims of too few cabinet posts for too many *Shimpotō* claimants. By the beginning of 1898, Itō was back as Prime Minister.

Itō tried to resurrect the politics of consensus by offering posts to both Itagaki and Ōkuma. It was all the more vital that he succeed, for by 1898 the Japanese Government was in some financial difficulty. Although revenues were steady and reliable, demands upon them were increasing steeply. In part, these demands arose from the need to prepare to retire the first of the bonds issued to finance the war with China. In part, they were required by Matsukata's programme for the creation of a set of public banks capable of marshalling private savings into long-term loans to promote agriculture, industry, and colonial development.[2] In even larger measure, though, the government's need was the military's, as both the army and navy proposed tripling their size. Such programmes required additional taxes and in Meiji Japan that meant land taxes. Landowners were particularly well represented in both Itagaki's and Ōkuma's parties.

Faced with this difficult situation Itō retreated to the familiar, inviting only Itagaki into his cabinet. But this step ran afoul of another *genrō*, Finance Minister Inoue Kaoru. So Itō submitted his land-tax proposals without any party's support. They were defeated roundly. Adding insult to this injury, the two opposition parties (*Jiyūtō* and *Shimpotō*) merged into one, the *Kenseitō* (Constitutional Party). Itō understood the implications perfectly well and devised a perfectly reasonable riposte: he would organize a political party of his own.

In the meantime, there remained the question of who would govern the country. No other *genrō* stepped forward, so Itō proposed another revolutionary development: that the *Kenseitō* leaders be tapped. Despite the *genrō*'s distaste for such a step – the first true party cabinet in Japanese history – there was little alternative. Itagaki and Ōkuma began to assemble a cabinet in June 1898.

That was a particularly delicate time for Japanese foreign policy. The situations of both Korea and China were quite fluid after 1895. Japan's victory over China had decisively separated Korea from that country. But Japan's influence over the Koreans was not assured as a result. The

Korean Queen Min was a rallying point for anti-Japanese Koreans. These forces could no longer look to China for support, so they turned instead to Russia. The Russians, for their part, were more than sympathetic. Their mammoth Trans-Siberian Railway project was, ideally, to terminate at a Pacific port that never froze over in winter. There were none such in Russian territory, but the southern tip of Korea had several excellent possibilities.

Needless to say, no Japanese leader welcomed a Russian-held port so close to the home islands. Unfortunately, the senior Japanese official in Korea after the Sino-Japanese War, General Miura Gorō, was exceptionally heavy-handed in protecting imperial interests. In October 1895, Queen Min was cruelly murdered and Miura was clearly involved. If Miura intended to destroy the anti-Japan forces in Korea, his actions badly backfired. The Korean King openly sought Russian protection and got it: in February 1896 Russian marines entered the Korean capital of Seoul. Pro-Japanese officials in the Korean Government were dismissed and many of Miura's Korean conspirators were executed.

These great setbacks revealed just how little the triumphs of the Sino-Japanese War had accomplished. If anything, they reinforced the *genrō*'s desire to avoid provocation while building up Japan's military strength (if the Diet would allow it by raising land taxes). But they also demanded a Japanese response. Yamagata (interestingly, and not a member of Itō's cabinet) took charge of that response. He journeyed to St Petersburg to negotiate the Yamagata–Lobanov protocol of May 1896. Technically, it was an even-handed document, providing for shared rights (specifically the right to station troops) in Korea for both nations. Actually, it did nothing to counter the growing Russian influence in Korea, though it did somewhat repair the damage to Japan's position in Korea caused by Miura's fumblings.

Matters were no better in China. Nor were they simpler. China had become the arena for competition and intrigue among the powers. By early 1898, this competition had become intense. Germany had been a late entrant into the race for empire and was determined to have concessions in China. As well, Germany hoped to frustrate growing cooperation between France and Russia in Europe by diverting Russian attention to East Asia. In November 1897, two German missionaries were conveniently murdered in China's Shantung province, prompting the German Government to demand a leasehold (in effect a colony) over Kiaochow, a city in Shantung. China appealed for Russian support in resisting this demand, but by early 1898 had to give in to German pressure.

What had happened to Russia's support? In essence, the Russians had reassessed their position in China. Although some Russian leaders favoured keeping China whole (and, therefore, resisting the German demands) so that all China would eventually fall under Russia's sway, those favouring a less ambitious policy prevailed in St Petersburg. Russia would permit Germany in Kiaochow in exchange for Russia obtaining Port Arthur at the southern tip of China's Manchurian provinces. Port Arthur could serve as a Pacific terminus for the Trans-Siberian Railway, obviating the need for a port in southern Korea. If no Korean port were needed, moreover, Russia might well be able to avoid a confrontation with Japan there.

The possibility of an understanding with Russia held great appeal for cautious Japanese leaders such as Itō, who was still (barely) in power as 1898 opened. Itō had his hands full in dampening the howls of outrage in the Japanese press over Russia's occupation of Port Arthur and of Germany taking in 1898 what Japan had been denied in 1895. Nevertheless, he pursued an understanding vigorously, proposing in January that Japan recognize Russian rights over Manchuria in exchange for Russian recognition of Japanese rights in Korea. But Japan's hand was not that strong. The resulting Nishi–Rosen agreement of April merely expressed Russo-Japanese recognition of Korea's independence and a pledge of non-interference in Korea's domestic affairs.

Even so, Russia had shown a desire for some sort of agreement with Japan, a sign of a slight recouping of Japan's position in Korea. In China, too, events of the spring of 1898 turned in Japan's favour with Great Britain's occupation of the port of Weihaiwei.

Japanese forces had been occupying Weihaiwei as a sort of collateral to ensure that China fully paid the Sino-Japanese war indemnity to Japan. The final instalment of the indemnity was delivered in the spring and Japan prepared to withdraw. At this juncture, the British approached Japan (and China) with a proposal to grant Britain leasehold rights to the port. It was not really Japan's right to grant Britain occupation rights, but that was just the point. The British approach to Tokyo before Beijing was a sign of a growing conviction in London that Japanese assistance might be useful against further Russian encroachments in China.

This British calculation eventually turned out to be correct. But in the spring of 1898 it was premature. Itō was much more interested in Korea than China and consequently much more interested in a deal with Russia, despite the rather disappointing Nishi–Rosen agreement. Nevertheless, he was willing to acquiesce in Britain's occupation of Weihaiwei so long as Britain recognized Japan's position in Fukien province (just across the Straits of Taiwan – which island was part of the Japanese Empire as a result of the war against China).

Things might have changed when Ōkuma became Prime Minister in June. But, as recounted above, Japan's internal politics were tumultuous, especially for Ōkuma who had nearly everything going against him. He

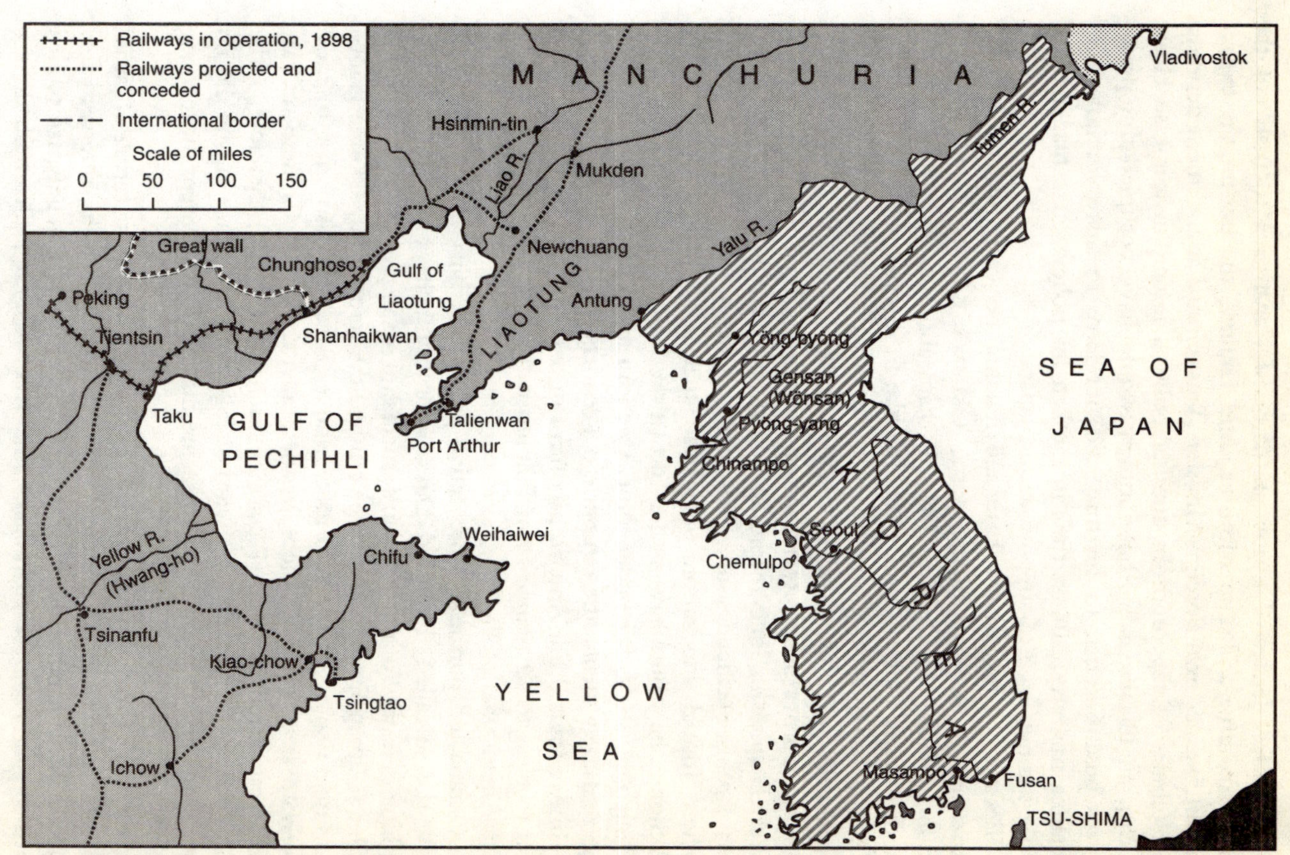

Map 2 Yellow Sea – Korea 1898

was the first leader of a political party (still anathema to many of the *genrō*) to serve as Prime Minister. And Ōkuma, the uncautious expansionist, was his own Foreign Minister. This unusual arrangement deprived his cabinet of support from another political leader, in this case Hoshi Tōru, an Itagaki man who normally would have occupied this prestigious post. Ōkuma, to be sure, did not have to share foreign-policy making this way, but by the same token his policy, once made, rested on a slimmer base of support within Japan.

Because of his narrow base and feuding between his and Itagaki's followers over government posts, Ōkuma's cabinet lasted barely 5 months, falling in October 1898. Perhaps also because of that narrowness, Ōkuma's foreign policy was Itō-like in its caution and moderation. Ōkuma had plenty of opportunities for bombast and bravado, but he remained quiet. The *Kenseitō*'s leader strictly followed his predecessor's policy of strict neutrality in the summertime's brief Spanish–American War. Despite his long record in public (and in publications) of favouring Japan's Pacific destiny, Ōkuma raised no objection to the United States' formal annexation of Hawaii that accompanied that conflict. Although he would have preferred that the former Spanish colony of the Philippine islands fall under joint American and Japanese administration, he remained silent when American President William McKinley began to move toward full American control over them. And he turned a deaf ear to pleas for assistance from Filipino nationalists who wanted to resist the Americans.

Ōkuma did pursue a more active policy in China. There, mid-1898 saw the so-called 'Hundred Days' reform, much of the effort led by Chinese eager to copy Japan's successful example of modernization. China's aging Empress Dowager crushed the reform, but Ōkuma granted asylum to some of the reformers who escaped. He also tried to move forward steps to bring more Japanese advisers to the Chinese military, steps calculated to lessen Russian influence there and, therefore, rather provocative. In this, Ōkuma was not following Itō's lead, but he did not last long enough in office to prosecute his anti-Russian steps more vigorously.

Ōkuma's October resignation was caused by a new fracturing of Japan's parties. Itagaki's supporters, led by the disgruntled Hoshi, reformed their new *Kenseitō* party without Ōkuma's followers, who spitefully named their new organization the *Kenseihontō* (the 'true' *Kenseitō*). This fresh division of the parties opened the way for a renewal of *genrō* influence. Suitably, the new Prime Minister was the arch-conservative Yamagata Aritomo.

Yamagata was conservative in two meanings of the term. At home, he rigorously limited the power of the parties even as he secured the *Kenseitō*'s collaboration with his cabinet. This feat was all the more remarkable since Yamagata's cabinet included no party members – and a full seven army and navy officers. In essence, Yamagata offered Itagaki and his followers an exchange. The politicians would not encroach upon the bureaucracy (especially the military) and its prerogatives. Indeed, in 1899 Yamagata would ensure that posts in the civil bureaucracy to as senior a position as Vice Minister could be filled only through Japan's examination system, not party appointment. The two military bureaucracies were, of course, even more insulated from party penetration – only active duty, senior officers would ever be allowed to serve in cabinets. The *Kenseitō* would help Yamagata's budget pass the Diet with a hefty increase in military spending funded by increased land taxes.

How did Yamagata accomplish this miracle? First, he opened the way to patronage jobs for the politicians and their friends. Japan's railways came under the state's control and embarked upon a colossal construction spree. In addition to providing jobs to build and then operate the tracks and stations, the railroads benefited the towns and villages that they reached. Even more directly, Yamagata bribed members of the *Kenseitō* for their votes, bribes worth nearly a million yen in all, a phenomenal sum at that time. He also consented to increasing the number of Japanese who could vote by lowering the wealth requirements for the franchise. And he agreed to permit secret ballots in elections. It was a curious alliance, Yamagata and the *Kenseitō*, but it worked. The politicians received money, patronage, and a broader reach into a bigger electorate. The conservative bureaucrats, especially the military, received bigger budgets, while retaining autonomy in the conduct of Japan's foreign relations.

Those bigger budgets allowed Yamagata to undertake a more expansive foreign policy than had been possible earlier. Although Yamagata retained a certain conservative sense of circumspection in his foreign policy, it could hardly be called passive. Yamagata was certain that the chief menace to Japan's security came from Russian designs in East Asia. It was a view seconded by his Foreign Minister, Aoki Shūzō, one of the first generation of real Japanese cosmopolitans (even in marriage: Aoki's wife was German). Yamagata and Aoki were particularly active in Korea, obtaining Japanese rights to build railways through much of the peninsula. It was not so much that Japan had an interest in trains (indeed, as Yamagata knew all too well, Japan barely had the financial resources to tend to domestic matters and its military budget). Rather, it was a strategy of forestalling: Japan had the rights so that Russia would not. As with any strategy, it was not completely successful. In the spring of 1900, Russia acquired a coaling base for its ships at Masampo, near the Tsushima Straits and, therefore, near Japan's home islands. Entering the twentieth century, Japan and Russia may not have been on a direct collision course, but tensions were evident and rising.

Those tensions rose further as a result of the Boxer rebellion in China. There, the forces of reaction were in power; the Chinese reform movement – a movement that had sought to imitate Japan's successful modernization – had been extinguished by the Empress Dowager. But while the Empress knew what she did not want, she had no easy alternative to meeting the challenge of the West. Their growing sense of helplessness and aimlessness led many Chinese to search for the source of China's troubles. They did not look far: foreigners had been harassed (and, in fairness, had been harassing in their turn) for years, especially since the rush for concessions in 1898. And these foreigners included the Japanese. Indeed, by 1900 many Chinese were looking at Japan not as a country to be imitated, but as an Asian society that had become corrupted by the West. Numerous secret Chinese societies, nativist movements, were organized to protect China against that foreign corruption. One of the largest of these was known to the West as the Boxers. In mid-June 1900, the Boxers were no longer even an open secret. Their followers entered Peking (with the Empress's tacit encouragement) to rid China's capital of its aliens, including the Japanese: on 11 June the Boxers murdered the senior clerk of Japan's legation in Peking in particularly brutal fashion.

News of the murder put Yamagata in a difficult position. There was little doubt that the Boxers' outrage would have to be avenged. But Yamagata coolly calculated that the result would be a large-scale military action against China by all the powers, probably leading to a new scramble for concessions, perhaps even the partition of China with much of it falling under Russian control.

Yamagata, therefore, sought to channel the anti-Boxer expedition into a limited effort, one that deliberately would not leave much room for Russian mischief. To this end, Foreign Minister Aoki approached Britain and Germany for support – against the Boxers and less directly, the Russians. Germany was lukewarm, keen on punishing the Boxers (who had murdered the German Minister in Peking soon after the Japanese clerk had come to his end), but also keen to keep the Russians deeply involved in Asia so that their alliance with France in Europe would mean little.

The British were an entirely different matter. They had long secured the dominant commercial position in China and so were reluctant to see China partitioned. They also had been engaged in imperial rivalry with Russia all along the rim of Asia, from Persia (now Iran) to India to China, and so had no desire to see Russian gains at China's expense. They were also in no position to help their cause in China. Britain was a naval power. What land power the British had was engaged in a protracted, dirty struggle against the Boers in southern Africa. The British, therefore, not only welcomed Japanese participation in the China expedition, they actively encouraged it, even offering to underwrite the costs of sending the Japanese forces.

Assured that Japan would not stand alone, Yamagata drew up plans that would see 22,000 Japanese troops sent against the Boxers, easily the largest contingent of any of the powers' forces. As importantly, he encountered no objections to the expedition within Japan. Only Itō cavilled, warning against becoming entangled in China's political and diplomatic morass, and against squandering Japan's still very thin resources on the expedition. Indeed, the Japanese public press urged Yamagata to use the Boxer crisis to consolidate Japan's position in Korea. But Yamagata aimed to contain Russian influence only, not confront Russia. He even agreed – to the great irritation of the Japanese public – to the appointment of a German to command the multinational expedition, all too aware that the Europeans and Americans (who had rushed a contingent from the Philippines) would not place their forces under a Japanese officer. Grateful for British support, he still feared a resurrection of the French–German–Russian combination against Japan of 5 years earlier. But he was determined that Japan should play its strongest role yet in influencing events within China and taking its place among the other imperial powers as an equal.

It is worth pointing out that Japan's foreign policy calculations, in 1900 as before, were made on the basis of the logic of empire and power. There was no sentiment in Tokyo to view the Chinese as fellow Asians suffering under a Western imperial system. There was scant desire to see the Meiji successes repeated by Chinese (or Korean) reformers. On the contrary, many thoughtful Japanese – who began to found various patriotic societies – increasingly believed that their own mere imitation of Western forms would never bring respect from other nations. The Europeans respected power, and Japan would have to acquire it at the expense of other Asians. One useful way to begin was the Boxer Expedition to Peking.

That multinational expedition had freed besieged Westerners in the Chinese capital by summer's end. The question then became one of determining China's fate. In many respects, the most important question – would China be partitioned? – had been settled even before the relief of Peking. The Japanese, the British, and the Americans had all indicated their opposition to partition while the fighting had still been going on (the latter going so far as to issue an 'Open Door' announcement favouring the preservation of China's territorial integrity). These powers had cooperated with the efforts of Chinese authorities in central and southern China to minimize Boxer disturbances there, and so foreign troops did not occupy those parts of China after the fighting in the north had ended. At a conference of the powers held after the relief of Peking, Britain successfully resisted setting compensation claims so high that China would have to pay in land (by granting colony-like concessions) rather than cash. Japan, still interested in obtaining additional finances, would have preferred a slightly higher indemnity from the Chinese, but strategic considerations outweighed financial ones, and Tokyo supported the British position.

But northern China, especially the northern-most provinces – those of Manchuria – was a different story. Whereas all of the other powers had landed their forces for the expedition to Peking from the sea, Russia had marched overland, through Manchuria. When the fighting ended, Russia had 50,000 troops in Manchuria, had occupied the key Manchurian port of Niuchuang, and had seized the railway bridging Tientsin and Peking. Moreover, the Russians had undertaken their own, direct negotiations with the beaten Chinese Government to embed Russia's paramountcy in north China into permanent, treaty form.

These developments were hardly welcome in Japan, where every leader had come to share Yamagata's desire to contain Russia in some fashion. The new-found cooperation with Britain and, to an extent, America, was welcome. But in the international system of imperialism in East Asia at the turn of the century, nearly everyone, not just the Japanese, saw Russia as the rising power, capable of bending the rules of that system to its benefit.

But there was substantial disagreement within Japan over what to do in response to this growing Russian threat. Again, the tumultuous internal politics of Japan played a large role. Yamagata had resigned as Prime Minister in September 1900. In part, he was irritated at Foreign Minister Aoki's direct report to the Emperor (that is, not reporting through the Prime Minister) of the recent events in China, a report Aoki had used to argue vigorously for stopping additional Russian encroachments, especially in Korea. Yamagata was concerned that Aoki might be playing to the jingoist press as a way to enhance his political future.

Primarily, though, Yamagata stepped down to force his old friend but current opponent Itō Hirobumi to assume power. This step may seem paradoxical – why force an opponent to assume power? But there was much logic in Yamagata's move. In Yamagata's eyes, Itō was committing the high sin of not only joining but actually forming a political party. Indeed, that was just what Itō had done. He had persuaded the *Kenseitō* to dissolve itself and form a new party, the *Seiyūkai*, and name himself as its leader. Itō had grand plans for the *Seiyūkai*, hoping to turn it into a coalition of political party leaders, businessmen – and bureaucrats. The possibility of this combination, especially breaching the walls of bureaucratic insulation, threatened the conservative structure that Yamagata had built. So Yamagata quickly resigned, before Itō had had a chance to recruit enough business leaders and especially bureaucrats into the *Seiyūkai*. As a result, Itō's cabinet had only politicians, old *Kenseitō* politicians, in it, except for the posts of army, navy, and foreign minister.

Yamagata moved to weaken Itō not only because he feared a threat to his domestic order, but also for foreign policy reasons. Prime Minister Itō was inclined to arrange an exchange with the Russians. If they would grant Japan a free hand in Korea, Japan would permit the Russians a similar status over Manchuria. Interestingly, Itō's own Foreign Minister, Katō Takaaki vigorously dissented. Katō had been an exceedingly curious choice for Itō's Foreign Minister since he was not personally loyal to Itō and was exceedingly ambitious, ready to play to the crowd patriotic spirit. He was an astoundingly young 41 years old. He had insisted on surprisingly stiff conditions for accepting the appointment: that foreign policy matters be considered in the Foreign Ministry only (a slap at the *genrō*), and that ministry officials not be automatically changed if a new foreign minister came to power, a further reinforcement of bureaucratic insulation above what even Yamagata had accomplished. Finally, Katō had served as Japan's Minister to Great Britain from 1895 to 1899 and was known by all to have pronounced anti-Russian views.

While it is a puzzle as to why Itō wanted Katō in his cabinet, it is no mystery that Katō wanted nothing to do with Itō's proposed Manchuria–Korea exchange. Katō wanted Korea, but he also had his eye on Manchuria. More immediately, that meant restraining Russia directly, not through trades. Katō understood that Japan itself was not strong enough for this task. But the completion of the Trans-Siberian Railway, now just a year or two away, would put Russia in a far stronger position then. And the recent Boxer affair had demonstrated that Britain might be a possible partner in restraining the Russians.

Katō lobbied Itō intensively, winning the Prime Minister's consent for a protest to Russia in the spring of 1901. The Russian reply, a brusque assertion that Japan had no right to meddle in Sino-Russian dealings, gave Katō the ammunition he needed to proceed toward confrontation. He continued his brisk protest to St Petersburg, and warned the Chinese that dire consequences would follow if they accepted any treaty arrangement with Russia. By April, Russia had withdrawn its treaty from China's consideration.

Shortly after, Itō's cabinet (the last to be headed by a *genrō*) fell. It was a victim not of Katōs diplomatic triumph, but its own political failings, as Yamagata had planned all along. Itō had never been able to enlist support from members of the House of Peers (the Diet's upper chamber), who held up his budget proposals constantly. Bureaucrats distrusted Itō's forming the *Seiyūkai* and refused to join it. In fact, they were angry over the *Seiyūkai*'s plans to increase government spending for patronage purposes beyond what the Finance Ministry thought was fiscally prudent – and without raising government (that is, the bureaucrats') salaries. Itō was out by May 1901, after only 7 months in office.

The new Prime Minister, Katsura Tarō, was the first of a new generation. Katsura had not been a player in the Meiji Restoration; he had been too young. Instead, he had risen, under Yamagata's supervision, through the new Imperial Japanese Army. Indeed, one might say that May 1901 was a milestone in the history of Japanese politics and foreign relations. Never again would a *genrō* be Prime Minister. And, for the first time, not only a bureaucrat but a military bureaucrat would hold that position. Katsura brought more than an army background into the prime ministership, he also brought the army's global viewpoint to the fore of Japan's foreign policy.

Katsura, like every other senior officer in the Imperial Army, believed that Russia was a mortal threat to Japan that had to be dealt with as rapidly as possible. But Katsura also shared his mentor's patience. Before confronting Russia, he would ensure that Japan would have an ally: Great Britain. Britain was a perfect choice for a number of reasons. The British and Russians had sparred against each other around the globe, so it was likely that London would be receptive to an alliance. Britain had the world's most powerful fleet, moreover, making the alliance popular with the Imperial Navy. Taking no chances on this score, Katsura had gone before the Diet to get substantial increases in warship construction funds.

Japan's Minister[3] in London, Hayashi Tadasu, opened the discussions for an alliance. Spurred by stories of new Russian encroachments in Manchuria, the British drew up a draft treaty by November, proposing in essence a defensive alliance. If either Britain or Japan found itself at war with one other power, its ally could remain neutral. But if either had to fight two (or more) enemies, the alliance would require the others to join the war.

These terms were most attractive to Katsura. He had never expected Britain to actively join in any war against Russia. But the alliance certainly would prevent any other European powers from joining with Russia, either during the war or after it, in a repeat of the Triple Intervention of 1895. Katsura, therefore, was delighted with the proposal, but Itō had other ideas. Itō had determined to travel overseas after his resignation as Prime Minister. Normally, this would have posed no complication. But Itō was still a *genrō*, still leader of the *Seiyūkai*, the majority party in the Diet's lower house, and he planned to see foreign officials overseas – including Russian officials. The possibilities for mischief were enormous. Fortunately for Katsura, the Russians took a very hard line with Itō. Itō had to return to Japan with no alternative to Katsura's British alliance as a way of dealing with Russia.

The Anglo-Japanese Alliance of 1902 was an important milestone in the history of Japan's foreign relations. Within 35 years of the Meiji Restoration, Japan had thrown off the unequal treaties. Its soldiers and sailors had defeated mammoth China and had marched alongside Western forces in China in the turmoil that had followed Japan's victory. Its leaders had secured a large empire, largely at China's expense, and a strong position in Korea. Now Japan had obtained recognition – in treaty form, no less – of its own great power status in an alliance-between-equals with one of the greatest nations of Europe. There was every reason for Japan to feel satisfied at its great progress by 1902.

There was also great reason to feel concern. Impressive as Japan's rise had been in East Asia, Russia's had been greater. The Russian rebuff to Itō, the occupation of Manchuria, the imminent completion of the Trans-Siberian Railway, indeed the creation of the Anglo-Japanese Alliance, all attested to growing Russian strength.

Katsura faced this strength from a weak political position at home. True, he had substantial support from the bureaucracies, both civil and military. Most of the *genrō* viewed him with favour. Thanks in no small part to Yamagata's influence, he also could count on the House of Peers as his preserve. But the Diet's lower chamber, the Commons, unfortunately had the power to stall, even veto, his budget proposals. By 1902 this hold-up was a serious matter. Katsura's own coalition, in part, depended upon securing additional funds for naval construction. And Katsura knew that he needed a strong army and navy to support a firm diplomatic stance against Russia. Katsura had hoped that his diplomatic triumph – the alliance – would break the *Seiyūkai*'s opposition to further increases in military and naval spending (he knew that the other party, the *Kenseihontō*, was an increasingly insignificant force in the Diet). But both parties blocked his budget, so Katsura called for Diet elections in August 1902.[4] The resulting Diet was no better, so Katsura dissolved it, calling for still another election in March 1903.

Katsura also had hoped that the Anglo-Japanese Alliance would lead to a more tractable Russia. Here, too, he was disappointed. Russia had made an agreement with China calling for a withdrawal of forces from Manchuria by April 1903. The time arrived and the Russian troops remained, prompting Katsura to meet with Yamagata, other *genrō*, and his cabinet's senior officials.

What emerged from these consultations was a not-so-new proposal: an exchange allowing Russia rights in Manchuria in exchange for Japan's rights in Korea. At first glance, it is difficult to determine why Katsura agreed to the idea, which had been rival Itō's for a long time and which the *Seiyūkai* leader embraced with relish. It is even more curious since the British, whom as allies Katsura had properly informed, were quite cool to it. And, to add enigma to the mystery, Katsura would offer to resign in July 1903 out of unhappiness with Itō's dual role as *genrō* and party leader, which Katsura thought gave his rival unfair political and policy-making advantages. So, why would Katsura agree to the 'exchange of rights' proposal to Russia?

The answer is to be found in the Japanese political situation. The elections of March 1903 had not diminished the *Seiyūkai*'s power in the Diet and Katsura very badly needed a budget. He therefore struck a deal with Itō, actually two deals. First, Katsura agreed to fund naval construction (and the budget increase it entailed) through a debt issue instead of new taxes, which would have fallen heavily on *Seiyūkai* supporters. Second, Katsura agreed to give Itō's policy towards Russia one more try.

But Katsura had by no means surrendered to Itō. He had outmanoeuvred him, in fact. His brief summertime resignation as Prime Minister, coupled with Yamagata's string-pulling, secured an imperial 'promotion' of Itō to the Privy Council. Itō would have to choose between remaining a *genrō* or the *Seiyūkai* party leader. He could not be both, since a partyman could not serve on the council. In reality, there was no choice at all: the Emperor had authorized the appointment, no loyal subject could refuse it. Itō had to resign from his *Seiyūkai*, a blow to his influence and to his former party's power.

Likewise in foreign policy, it seems that Katsura had the better estimate of the international situation. He must have known that Britain would not welcome a Manchuria–Korea exchange. He likely suspected that Russia would not agree either. In that suspicion, he had already got Itō's confirmation that Japan under no circumstances would agree to concessions to Russia in Korea.

Yet that was exactly what Russia would demand. By the summer of 1903, the short cut of the Trans-Siberian Railway, across Chinese territory to the Russian Pacific port of Vladivostok, was complete. Already, the Russians had plans for a railroad from the northern border of Korea, the Yalu River, to the Korean capital of Seoul, a direct strike at Japan's influence throughout the northern portion of the country. The Imperial Army closely monitored the construction of the railway, and was well aware of Russian railroad survey teams visiting Korea. Katsura, then, could not have been greatly surprised by Russia's September reply to Japan's proposed exchange: Japan could exercise rights in Korea south of the thirty-ninth parallel. North of it, Korea would be neutral, under no power's sway. About Manchuria, the Russians said nothing, and hence expected no limits on their activities there. While awaiting the Japanese reply, Russia improved the fortifications of Niuchuang and Port Arthur, basing battleships there for the first time.

Katsura, already armed with a decision not to back down over Korea, sent Japan's counter in late October. Japan would agree to a neutral zone, but along the Yalu River, effectively giving all Korea over to Japan's interests. Russia replied by mid-December with a blunt rejection. At the same time, Russian troops occupied the Manchurian city of Mukden. Almost immediately the Japanese Diet was filled with howls of protest against Katsura for failing to stand up to the Russians.

It is possible – and correct – to see these protests as one more indication of how little the Diet mattered in the formation of Imperial Japan's foreign policy. Diet members were completely in the dark about the real thrust of Katsura's plans. They could interpolate the Prime Minister and his cabinet, but the ministers were not obliged to reveal information that might compromise Japan's security. They could pass budgets for the army and navy, but had little idea of what the services did with the money or to what purpose. It is worth remembering, moreover, that the political parties were generally satisfied with this arrangement. They would leave the bureaucrats, civil and military alike, to run the nation, so long as ample patronage jobs (whether at a railroad station or army depot) and cash were available for serving their constituents and their re-election campaigns.

Of course, well before these Diet protests, Katsura had moved to stand up to Russia. He already had the support of the Imperial Army's General Staff, which was alarmed at the speed of the Russian military build-up in East Asia. The navy, too, was concerned at Russia's sharply increased naval presence in nearby waters. So the players that mattered – the army, the navy, and the civilian ministries – supported Katsura. Only the cautious *genrō* held back, at Itō's urgings. Itō was strongly opposed to war against Russia. But even Yamagata wanted to be sure that hostilities were used only after all other means to adjust relations had been attempted – though he was gravely disturbed at Russia's recent penetrations deep into Korea.

So Katsura sent a final proposal to Russia. It was something of an anticlimax, essentially unrevised from the earlier one offering the Yalu as a demarcation between Russia's sphere in Manchuria and Japan's in Korea. The real substance of the December decisions of the cabinet, general staffs, and *genrō* authorized Japan's preparations for war in case (though it might be better to say 'when') the Russians refused this final offer. Japan had to preserve the support of Britain and the United States during any conflict with Russia. Both of these powers strongly opposed any partition of China. Therefore, China would have to be kept neutral during any war to avoid its being on the losing side and vulnerable to partition. In the meantime, the Imperial Army began preliminary mobilization while the navy rushed to arrange sea transport for the troops and to acquire extra warships through purchase from European yards.

The Russians were not hasty in responding to this last Japanese proposal. It contained nothing new. And the Russians found it difficult to believe that Japan would accept war with the dominant and still-rising power in East Asia. They were wrong. On 8 February 1904 the Imperial Navy attacked Port Arthur and blockaded it. Simultaneously, Japanese troops deployed to Inchon, Korea. Two days later, Japan declared war.[5]

The Russo-Japanese War went well for Japan. Korea was secured virtually from its outset. China's neutrality was easily obtained. Japanese troops landed on the Liaotung peninsula in early September, sealing off Port Arthur from land. The ensuing siege of Port Arthur prevented any Russian seapower from menacing Japanese operations. The port itself fell on New Year's Day, 1905. Over half a million troops fought in the climactic, and bloody, battle of Mukden that drove the Russians completely out of southern Manchuria by mid-March.

The Japanese public was nearly delirious over the victory at Mukden. But cooler leaders, such as Yamagata and Army Minister Terauchi Masatake, better understood the implications of that battle. That Russia had been able to muster such a large number of troops in the Far East was clear evidence of its rapidly rising ability to reinforce itself for still more, and still larger, conflict. While Japan certainly had additional troops available, raising them would cost time and money. Money could be had: Japan had experienced no difficulties raising it through bond issues overseas, but it was disconcerting to have to rely upon foreign capital. In any event, battlefield losses already were high. It was time to end the war, therefore, at the moment of Japan's greatest triumph and before Russia could strike back. On 21 April 1905, the cabinet agreed upon peace terms.

But Japan did not yet call for peace. There were two immediate difficulties. One was the Russian Baltic fleet, which had departed European waters in a colossal global voyage (made still longer by Britain's refusal to permit the use of the Suez canal) to reach Vladivostok. On 30 May, it reached the Tsushima Straits between Korea and Japan, where it was intercepted by the Imperial Navy. The resulting engagement destroyed the Russian fleet and provided the perfect moment to move toward peace.

The second difficulty, of course, was to move toward peace without seeming weak. There could be no suspicion that Japan indeed was running out of money – else Russia would either refuse to negotiate an end to the war altogether or simply prolong peace negotiations to prepare for its revenge. Fortunately, the United States had a particularly dynamic

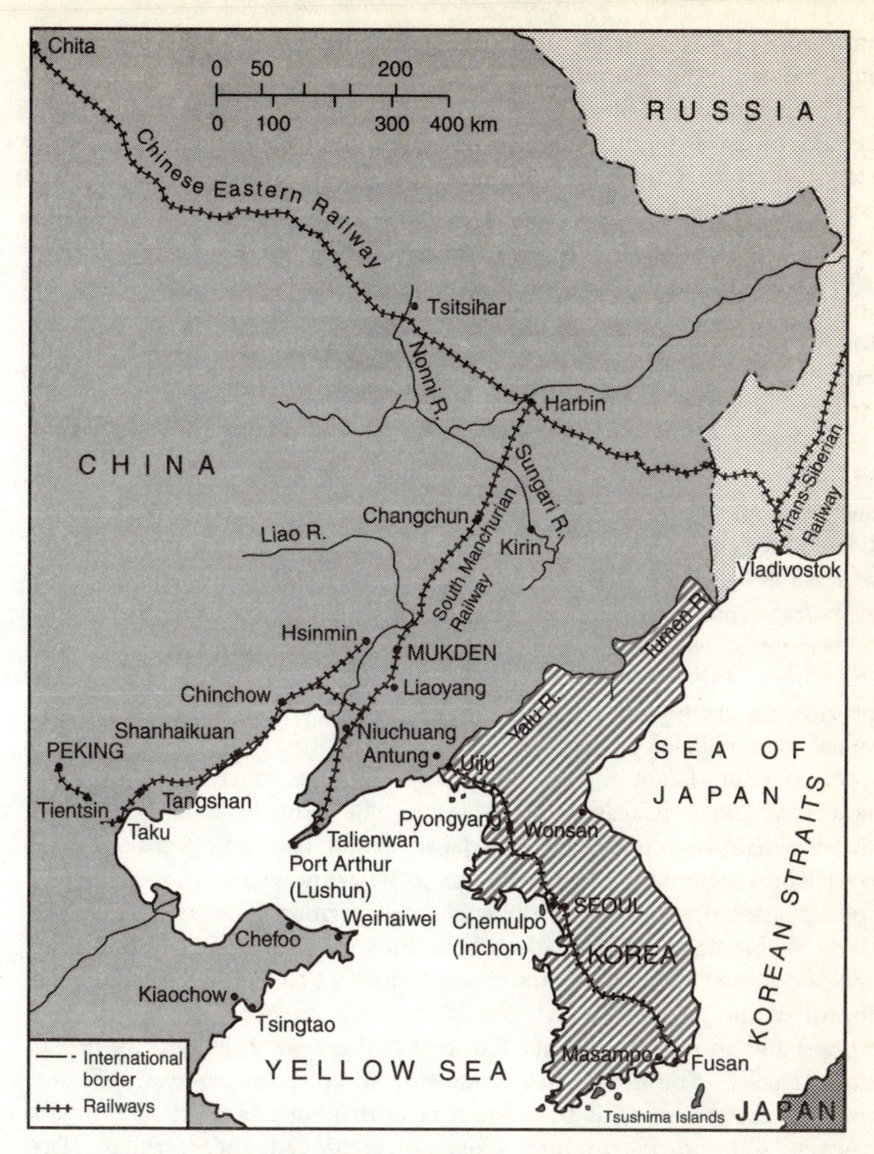

Map 3 The theatre of operations, Russo-Japanese War 1904–5

president at this time: Theodore Roosevelt. Roosevelt, concerned that neither Russia nor Japan secure too great a dominance over Northeast Asia, was willing to make any peace initiative appear as his own, not Japan's. On 1 June 1905, Japan privately requested America's 'good offices' to negotiate an end to the war. In Russia, too, financial considerations and an increasingly unhappy public argued for peace. And so in early August Russian and Japanese representatives met with Roosevelt at Portsmouth, New Hampshire.

The question in Tokyo was who exactly would represent Japan. It was certain to be a highly unwelcome task, not because difficulties were expected from Russia. No, the problem was that there simply was no way that any formal peace terms could live up to the unbounded expectations of the Japanese people. For this reason among others, Katsura wanted a *genrō*, preferably Itō, to take the job. But Itō had no desire for such duties, especially in the service of Yamagata's protégé. Yamagata, for his part, did not want the blame for the treaty to fall upon any of his cronies, the so-called 'Chōshū clique', and so the task fell to Foreign Minister Komura Jutarō, who came from an obscure clan in Kyushu. The military, however, feared that Komura would pursue a very hard line toward the Russians, because Komura had done so before the war and he would not want the opprobrium of 'soft' peace terms in any treaty he would bring back. In fact, military leaders – the men most directly and painfully aware that the war should not continue – suspected that Komura would break off peace talks rather than accept a treaty that would be unpopular with the Japanese people. In the resulting struggle between the army and the Foreign Minister for control over the terms of the peace process, the army won, a further sign of the rising influence of the professional military over Japan's foreign relations.

As a result, Komura was subject to a number of restraints over his powers as senior delegate to Portsmouth. The terms he took there stated that the main objective was peace. Japan would insist on effective control over Korea. Southern Manchuria was to be acknowledged as territory of special interest, with Japan inheriting all former Russian concessions there, including the prized South Manchurian Railway. But nothing else was 'absolutely essential'. Komura was to attempt to secure Sakhalin island (north of the Japanese home islands, it had been occupied by Japanese troops) and an indemnity from Russia, but these were desires, not necessities. Finally, Komura was to consult with Tokyo in every detail, and not take matters into his own hands at Portsmouth.

Even so, the Portsmouth Conference proved quite trying. The Russians easily granted Japan's essential terms but objected to losing Sakhalin and were firmly set against any indemnity. Komura was just as firmly set upon gaining at least some territory and a measure of monetary compensation. He proposed that Japan technically drop its demand for indemnity, but give the northern half of Sakhalin back to Russia in exchange for a payment of over 1 billion yen.

It was an ingenious plan, but neither the Russians nor Roosevelt accepted it. Komura then suggested to his superiors in Tokyo that he leave Portsmouth and draw up a declaration blaming Russia for the talks' collapse. The army and navy immediately vetoed this idea, and Komura was instructed to get a peace treaty even if neither Sakhalin nor an indemnity could be obtained. A bitter Komura held out on his own, and did win the southern half of Sakhalin for Japan, but no money. On that basis, the Portsmouth treaty was signed on 5 September 1905.

By then, word of its terms had leaked out. The result was a series of demonstrations-turned-riots, the biggest in Tokyo's Hibiya Park. Many Japanese had fantastically expected their country to gain most of Russian Siberia, or at least its Pacific coast. Komura was widely vilified. Katsura, hardly popular himself, won no friends among the public with his decision to suppress the riots quickly and directly. But that was just the point. Katsura did not need public support to see his peace programme through. The Privy Council ratified the treaty without difficulty. The Diet was another matter. There, the *Kenseihontō* plotted to ally with the *Seiyūkai* to form a unified opposition against Katsura. But its members failed to reckon with the Prime Minister and one of the ablest politicians of his generation, the *Seiyūkai*'s new leader, Hara Kei.[6]

Hara had risen through his determination and ability to use the new institutions of Meiji Japan. Born in the remote northeast, the young Hara had tried to enter the Meiji bureaucracy without success, failing entrance examinations for both the navy and foreign ministry. But he persuaded *genrō* Inoue Kaoru, and then Mutsu Munemitsu, to sponsor him in a successful career in the Foreign Ministry, where Hara became Vice-Minister (under Mutsu) in 1895. After fifteen years as a bureaucrat, Hara dabbled in journalism for a few years before joining the *Seiyūkai* soon after its founding. His supervision of much of the party's finances made the party a force to be reckoned with in the Diet, and Hara a force in the party. When Saionji Kimmochi succeeded Itō as party leader, Hara became Saionji's right-hand man (in fact, some would say Hara was more head than hand of the *Seiyūkai*'s nominal chief).

When Prime Minister Katsura pondered how to avoid a coalition against him in the Diet, he approached Hara. Would the *Seiyūkai* support his programmes, including the retention of wartime taxes and military-rich budgets, and avoid criticizing the Portsmouth treaty? In exchange, Katsura promised his own early resignation, with Saionji to become the next Prime Minister, so long as not all cabinet posts were filled with *Seiyūkai* members. Hara agreed. In January 1906 the Saionji Cabinet was formed, with Hara as Home Minister.

For a time, the Saionji Cabinet had the luxury of few problems in Japan's foreign relations. Before resigning, Katsura had overseen the renewal of the Anglo-Japanese Alliance. It was useful insurance against the possibility of a Russian war for revenge, as it now stipulated that its provisions would come into play if either power were attacked by a single other country, not just two or more as previously agreed. And it included British recognition of Japan's position in Korea, a virtual *carte blanche* for Japan to move toward annexing that country. Katsura had also initiated discussions with Russia, much more exhausted itself by the war than Japan had realized, that would by early 1907 lead to an agreement partitioning Manchuria into Russian and Japanese spheres. Russia, too, gave Japan a free hand in Korea (in exchange for identical rights in outer Mongolia). Japan made Korea a protectorate in November 1905. A month later, Japan secured China's acknowledgement that former Russian rights in Manchuria were now Japanese. Moreover, all the major powers upgraded the status of their diplomatic missions to Japan from legations to embassies, a symbolic recognition of Japan's new status as a major power.

Surprisingly, Japan's chief difficulty came from the United States. The initial cause, perhaps as surprisingly, was not Japan's acquisitions on the Asian continent. The Taft–Katsura agreement of July 1905 served as America's recognition of Japan's position in Korea and Manchuria. But the next year saw the United States Congress pass legislation barring Japanese immigration, and the city of San Francisco acted to segregate Japanese (and other Oriental) students from whites in its schools. Both actions were keen reminders to the Japanese that, despite their accomplishments in diplomacy and on the battlefield, many still viewed them in terms of their race, a race still viewed as second-class.

There were hotheads on both sides of the Pacific. Some in America called for boycotts of Japanese businesses (perhaps the first instance of 'Japan-bashing' in the United States); others fanned rumours of secret Japanese spy rings and war plans to wrest America's Pacific possessions away and called their fellow citizens to arms. In Japan, there were the inevitable cries for resistance to the American outrages. And a former naval officer published a book about a future war with America that, sure enough, wrested its Pacific possessions away.

But both countries had leadership interested in reducing friction. Saionji's Foreign Minister, Hayashi Tadasu, suggested that Japan voluntarily restrict emigration to the United States to eliminate the need for America's immigration and segregation laws. President Theodore Roosevelt, an open admirer of Japan's accomplishments, embraced the idea. It would become the 'Gentlemen's Agreement' of December 1907. In fact, Saionji's mentor, Itō, had recommended going even further to accommodate the United States: an explicit understanding with Washington that the United States had complete freedom to regulate its immigration questions without any outside interference.

Even Roosevelt's ostensibly provocative act, sending the American battle fleet on a world cruise, turned out to have a palliative effect.

Roosevelt had several purposes in mind when the idea for such a tour germinated in late 1907. He wanted to demonstrate to alarmists along America's Pacific coast that there was no threat to their security. Along similar lines, he wanted to determine if the fleet was capable of long-range steaming. And the cruise would generate publicity favourable to increased naval spending, one of Roosevelt's top priorities as a president dedicated to securing for the United States a larger role in global affairs.

Somewhat to Roosevelt's surprise, the Saionji Government asked that the American battleships stop over at Yokohama. They arrived in October 1908, several months after Katsura had returned to power, but Saionji's objective of a tangible sign of reconciliation was achieved splendidly. Thousands of American sailors received an exceptionally warm reception in Japan. Taking advantage of the new mood, Komura, back as Foreign Minister, had his ambassador to Washington meet with the American Secretary of State to discuss outstanding differences. The resulting Takahira–Root agreement of November pledged both countries to increased commerce, respect for the Open Door in China, and reciprocal respect for each other's Pacific possessions.

This agreement was entirely in keeping with Katsura's determination to make real the so-far largely 'paper' rights Japan had won in southern Manchuria. But that determination led to a new collision with a new American administration, that of President Robert Taft and his acerbic Secretary of State, Philander Knox. Knox was concerned that Japan's control of the South Manchurian Railway, a line that ran from north of Mukden directly south to the Manchurian ports, gave Japan an economic stranglehold on the region – to the detriment of American business – and a political grip on it as well – to the detriment of Chinese territorial integrity there. Knox grasped at various ways to weaken Japan's position. One involved building a parallel and, therefore, competitive line from Chinchow to Aigun with American and British financial assistance. Another, grander proposal was to float an international loan to China to purchase all foreign-operated railroads (including the South Manchurian Railway), thus 'neutralizing' Manchuria from foreign influences.

None of Knox's schemes succeeded, but these American initiatives did have consequences for Japan's foreign relations. Britain's Pacific dominions, Australia, New Zealand and Canada, grew increasingly concerned with the possibility of an American–Japanese clash, with Britain still allied with Japan. These countries operated immigration policies similar to America's, and they feared that, if Japan should prove menacing in the future, they would have to look to Washington, not London, for possible protection. (Knox had proposed to the British a common policy on immigration issues, much to London's annoyance.) The tangible result was a reworked Anglo-Japanese Alliance in July 1911 with a provision that neither signatory was obliged to go to war with a nation it had signed an arbitration treaty with. (Britain was on the verge of signing such a treaty with the United States. Ironically, the American Senate refused to ratify the treaty, throwing much of this diplomatic work overboard, to London's frustration.)

More importantly, Knox's futile efforts were a great help to Katsura's attempts to solidify further Japan's position in Manchuria. The American's neutralization scheme had threatened not just Japan's railroads and influence, but also Russia's. The predictable result was a series of Russian–Japanese discussions leading to an agreement promising mutual recognition of spheres of influence in Manchuria signed in January 1910. Far from prying loose Japan's hold on southern Manchuria, Knox's blunderings had strengthened it.

Soon after the agreement with Russia over Manchuria, Japan annexed Korea. In part, this momentous step was made possible by international consent. Russia and Britain had clearly indicated their tolerance for annexation. The United States had given approval by means of the Taft–Katsura agreement. The other powers had not been major factors. But the annexation itself was triggered primarily by the tragic events in Korea after the Russo-Japanese War.

There was no question that Korea was Japan's preserve after that war. But there was much debate over the type of preserve it was to be. Annexationists, who preferred direct Japanese rule over Korea, were disappointed with the selection of Itō as Korea's postwar Governor-General. Itō, one of the giants of the Meiji Restoration, hoped to bring a similar miracle of modernization to Korea. To this end, he insisted on ruling with Korean assistants, and it fair to say that some elements in Korea that had always looked to Japan as a model of modernization flocked to Itō's side. Itō also successfully demanded control over all Japanese activities in Korea, including the Imperial Army. The army bitterly complained about this infringement on its autonomy, but gave in.

Itō's efforts failed, but not because of the Imperial Japanese Army. The Korean Emperor had remained in place after the Russo-Japanese War. He was not a figurehead; he held real power and had many followers opposed to modernization. In fact, Koreans favouring reform took the initiative in persuading Itō to press for the Emperor's abdication in the summer of 1907. This step, coupled with Itō's disbanding of the Korean army, led to widespread riots suppressed by the only tool Itō had: the Imperial Japanese Army. At least 10,000 Koreans were killed, further embittering Korean nationalists. They scored revenge at a railway station in Harbin, Manchuria, assassinating Itō in October 1909. In this way,

Itō, who had cheated death and spearheaded reform with immense success in Japan since the 1860s, met his end in Korea. If the Koreans hoped to achieve independence by the murder, they were disappointed. With Itō removed, there were no obstacles in Japan to proceeding at once with formal annexation. The new Governor-General was an army man, Terauchi Masatake, and he immediately began a harsh repression of all those opposed to Japan's rule.

By the start of 1911, Japan had resolved nearly all of the outstanding issues in its foreign relations. Tokyo had widespread international recognition for its position in southern Manchuria. Korea had been annexed. And, at long last, the final provisions of the old unequal treaties restricting Japan's ability to set tariffs at any level had been swept away. Japan was not only a fully sovereign nation in 1911. It was also a true empire in East Asia, one that rivalled all the others and had bested in combat the most menacing. In recognition of his role in these achievements, Katsura was given the title of Prince in the spring.

But there were reasons for concern, too. The first was financial. Shortly after the Russo-Japanese War, Great Britain's Royal Navy had introduced a revolutionary new type of battleship, the dreadnought. In a stroke, it made all the older battleships in all the world's navies obsolete. If the Imperial Navy was to retain its hard-won standing in the world, therefore, tremendous new expenditures were necessary. Increased naval expenditures posed two additional problems for Prime Minister Katsura. The army, his political power base, would hardly tolerate a huge naval expansion without securing additional funds itself. But the *Seiyūkai*, under Saionji and Hara, was increasingly insistent that the tax burden, and, therefore, government spending, be reduced. Also, as the renewal of the Anglo-Japanese Alliance demonstrated, racial issues were a growing irritant in Japan's foreign relations, making its leaders all too aware of their nation's potential isolation despite its past successes. Japan had an empire, but it was still not clear that the empire ensured Japan's security.

Of all these problems, the most immediate was the budget. Rather than make a choice among the navy, the army, and the *Seiyūkai*, Katsura resigned in August, handing over the nettle to Saionji. The new Prime Minister barely had time to consider Japan's finances when the Chinese Revolution broke out in October, presenting Japan with bold new opportunities for imperial expansion. The debates within Japan over how best to seize these opportunities would help lead to deep internal splits in the old Meiji hierarchy and a new round of confrontations with the West, itself about to be shaken to its foundations by the First World War. Japan's search for security through empire was about to encounter new difficulties.

Notes

[1] To vote, one had to be at least 25 years old, a male, and had to pay a land tax – meaning that one had to own land. Barely over 1 per cent of the population qualified.

[2] These closing years of the nineteenth century would see the creation of the Industrial Bank of Japan, the Hypothec Bank of Japan, and the Bank of Taiwan, for example.

[3] At this point, Japan was still not accorded recognition as a power of the first rank, so its top diplomats abroad were titled ministers, not ambassadors.

[4] It is important to remember that from 1868 to 1945, the cabinet was formally appointed by the Emperor (and actually selected by a group of senior leaders, at first the *genrō* and later a collection of retired prime ministers). Only members of the Diet were subjected to the electoral contests.

[5] There has been some confusion among scholars over the timing and motives of Japan's attack on Port Arthur. Some have argued that it presaged a similar treachery to the Pearl Harbor attack against the United States in 1941. This is misleading. Japan had broken diplomatic relations with Russia before the attack and considered the break as suitable warning to Russia.

[6] Hara's given name can also be romanized as 'Takashi'.

4
The rules change

By 1911, Japan appeared to have both freedom and security. But that year ushered in the first of four severe challenges for Japan. The Chinese Revolution shattered even the semblance of stability on the Asian continent. It was impossible to tell what might emerge from the ruins of China's decrepit empire. The stakes were high for Japan, especially in Manchuria. Three years later, the outbreak of war in Europe began the end of the European empires themselves, adding more uncertainty to how Asia might be reordered and who might do the reordering. In 1917, the Russian Revolution provided one alternative: how far would communism spread through Asia, in Siberia to the north or China in the centre? And in the United States, President Woodrow Wilson offered another alternative for the new rules of Asia's international relations: an end to secret deals at others' expense and respect above all for the right of all peoples to determine for themselves their political systems free from foreigners' interference.

These four challenges would have been difficult enough to meet if Japan had had a clear response to each of them. But increasingly Japan did not speak with one voice during these years. With age, the *genrō*'s voices weakened, permitting new élites to bid for control over Japan's foreign policy. Chief among these new élites were the bureaucrats and the politicians. It would be a mistake to regard either as monolithic. The bureaucrats in various ministries often disagreed with each other. Increasingly the army and navy saw themselves as apart from the other ministries. Likewise, the politicians were openly divided into contending parties which themselves were often split internally. The politicians, though, did wield more influence as the years progressed, primarily through their growing association with a new élite: the Japanese business community, especially four colossal industrial conglomerates,[1] Mitsui, Mitsubishi, Sumitomo, and Yasuda, called *zaibatsu* (or 'financial groups'). These combines often funded parties, forming alliances with them, in order to secure legislation favourable to their interests. It was a successful and profitable arrangement, although it usually did not affect Japanese foreign policy.

None of these new élites succeeded in gaining control of that foreign policy, in fact. As a result, Japan – or rather, various Japanese leaders – often pursued different and sometimes contradictory policies. At first, these differences were reparable. In the end, by 1941, they brought catastrophe.

Some of these differences emerged days after rebellion came to China. Despite decades of trouble there, the Chinese Revolution of October 1911 still took Japan by surprise. As the Saionji Cabinet scrambled to formulate a response, it was guided by three general objectives. Japan's position in Manchuria, centred around control of Dairen and the South Manchurian Railway, was to be maintained at all costs; the gains of the Russo-Japanese War could not be sacrificed. Japan would encourage the emergence of a Chinese regime friendly to further Japanese penetration of the mainland. And Japan would attempt to preserve the institution of monarchy – the emperorship – in China.

Unhappily, these three objectives did not mix well. The *genrō* strongly opposed anything that might aid the creation of a Chinese republic, even one friendly to Japan and Japan's interests in Manchuria. Their stance was unsurprising. They had dedicated their lives to the restoration of an imperial institution to Japan and were not about to assist in the overthrow of an emperor in China. Perhaps even more importantly, they had no wish for developments in China to encourage radical republican movements in Japan.

Saionji's Foreign Minister, Uchida Yasuya, agreed and recommended military aid for the besieged Manchu regime, though he wanted recognition of Japan's position in Manchuria in exchange. Rising *Seiyūkai* star Hara Kei favoured aiding both Manchus and rebels. To him, the most important thing over the long term was a China friendly toward Japan. The Imperial Army disagreed. It focused on the possibility of another war against Russia. As a result, the army valued Manchuria foremost and saw the overthrow of the Manchu Emperor as the best way to secure actual independence of Manchuria from China, so it argued for aid only to the rebels. China's friendship was nice, but hardly necessary, and if the choice were between a friendly China and a Manchuria securely under Japan's control, the army's choice was obvious.

Events disappointed everyone. The Manchus recalled General Yuan Shih-k'ai to service to suppress the rebels. Yuan had a long record of anti-Japanese activity, going back to his actions in Korea before the Sino-Japanese War, so his new role displeased Tokyo. Upon his return to the Chinese capital, Yuan turned to Great Britain for support in negotiating a truce with the rebels. Once it was achieved, Yuan turned on the Manchus, eased them out of power, and had himself confirmed as the first President of China's new republic in February 1912.

For Saionji, Yuan's timing was abysmal. Saionji had come to power the previous summer determined to restore discipline to Japan's finances. This stance pleased fiscal conservatives. But it was unpopular with those who, on the contrary, actually wanted the government to spend more money. Hara wanted bigger budgets to keep the *Seiyūkai*'s patronage machine operating smoothly. More importantly, both army and navy demanded large new spending increases to meet the uncertain new international situation. Meanwhile, Saionji's old political rival , Katsura Tarō,[2] manoeuvred to establish a broad-based political party of his own, the *Dōshikai.*

As the Diet opened in early 1912, all these groups used the China issue to pillory the hapless Saionji. They drew a picture of a prime minister without a foreign policy, or at least without an effective one. Saionji gamely weathered the attack and prepared to face down his opponents in the Diet at the end of the year. In this, he was aided by Katsura's inability to attract many Dietmen to the *Dōshikai* and by the Emperor Meiji's death in July 1912, which occasioned a brief political truce.

But Saionji made a crucial misstep as he prepared for a new confrontation in late 1912. The budget Saionji would submit to the Diet indeed cut back on spending in every category but one: the Imperial Navy would receive its requested increase, a substantial one.[3] The Imperial Army was outraged. Normally, Saionji would have turned to a senior political figure with army ties to sooth feelings. Normally, Katsura would have been the ideal candidate. But Katsura did not want to help Saionji. He was caught up in the formation of his *Dōshikai* party. And, because he was forming a political party, he had lost favour in the eyes of Yamagata and the army. Both the army and Yamagata were also angry that Katsura had not persuaded Saionji to send reinforcements to Manchuria upon the outbreak of the Chinese Revolution.

Instead of mediation, therefore, there was crisis. In November 1912 the army withdrew its minister from Saionji's cabinet, which immediately fell since the army refused to name a replacement. The disgruntled *genrō* realized that any successor had to be someone with support from the Diet and the bureaucrats and be at least somewhat acceptable to the army. The only person appropriate was Katsura, who became Prime Minister in December.

Even so, Katsura was hardly in a strong position. In fact, his becoming Prime Minister illustrated how splintered power had become in Japan as *genrō* influence began fading. His *Dōshikai* was no match for the *Seiyūkai* in any Diet contest. His relations with the army, though far better than Saionji's, were still strained. To remedy this first difficulty, Katsura moved to call new Diet elections. But before he could even hold these, the *Seiyūkai* called for his resignation by moving a vote of no-confidence in Katsura's government.

Katsura, finding the prime ministership too weak for this challenge, turned to the newly installed Emperor for help. In a manner reminiscent of the early days of the Meiji Restoration, Katsura secured an Imperial Rescript commanding the *Seiyūkai* to call off its upcoming vote. Forty years earlier, it would have worked. But, to Katsura's amazement and chagrin, the *Seiyūkai* defied the rescript. Mass demonstrations blasted Katsura for his heavy-handed, 'undemocratic' tactics and, most of all, for his subjecting the Emperor to embarrassment.

At this point, Japan might have slid into political paralysis or even civil war. Katsura might have called upon the Imperial Army to enforce the Emperor's will (even though the Emperor actually had simply rubber-stamped Katsura's decision). But Katsura cared too much for the Emperor to risk that. Badly wounded by the charges that he had embarrassed the Emperor by securing a rescript that was disobeyed, a stunned Katsura resigned the prime ministership and died within months.

The Imperial Army did not escape this 'Taishō Crisis' unscathed, either. The victorious *Seiyūkai* allied with the navy to install Admiral Yamamoto Gonnohyōe as the new premier. To ensure that the army did not sabotage the Yamamoto Cabinet by refusing to provide a minister, Yamamoto ruled that generals *not* on active duty were eligible to serve in that capacity. In this way, he (or any other prime minister) could appoint an army minister from retired officers who, precisely because they were retired, were not subject to the army's orders that they refuse to join any cabinet. The army protested, but had little alternative but to go along, still without its increased budget. A sullen army retreated, waiting for a chance to strike back.

That chance came within a year. In January 1914, news broke that the Imperial Navy had accepted bribes in its awarding of the new construction contracts. The *Dōshikai* and the army instantly attacked Yamamoto and his *Seiyūkai* allies in the Diet and blocked additional naval spending. Yamamoto fell in March and the *Dōshikai* took power.

Throughout 1912 and 1913, Japan's foreign policy drifted while furious political battles were being fought at home. But foreign concerns reclaimed centre stage with the outbreak of war in Europe by August 1914.

War in Europe offered Japan a precious opportunity for gains in Asia. It was determined to move swiftly and decisively to secure those gains. This opportunity was immediately recognized by the *Dōshikai*'s leader, Foreign Minister Katō Takaaki. Katō was an exceptionally ambitious man, with very strong views on Japan's foreign relations. He believed that Japan had no time to waste in enlarging its position on the Asian continent at China's expense. In his mind, Japan was the future leader of all East Asia, and the quicker China could be brushed aside the better.

The best place to start the 'brushing' was Germany's set of holdings and concessions throughout East Asia, but especially inside China. Germany held colonial-type rights over the strategic and heavily populated province of Shantung. In addition, the Germans owned several island chains in the Pacific, such as the Solomons, Marshalls, and Carolines, which commanded important approaches to Japan from the ocean.

When Britain declared war against Germany in early August, both European powers hoped to avoid the spread of hostilities to Asia and the Pacific. But London soon concluded that German commerce raiders there were doing sufficient damage to British shipping to require Japanese naval patrols under the terms of the Anglo-Japanese Alliance. It was Britain's bad luck that its request for help went to the ambitious Katō.

Japan's intervention in the First World War was not automatic. Some leaders, especially within the army, believed that Germany would win and that to alienate Berlin, therefore, was unwise. It was better to watch and wait. Katō overruled this caution. He was confident that Britain would not be on a losing side. And, in any event, he was eager to expand further into China. It would be foolhardy to stand by while the great prize of Shantung awaited. Sweeping aside British reservations and an American proposal to 'neutralize' the Pacific, he won the cabinet's consent to dispatch an ultimatum to Germany on 17 August.

This ultimatum predictably angered Germany, which just as predictably rejected it. But it also horrified China and Britain, because Katō demanded the surrender of Germany's Kiaochow leasehold in Shantung. Japan, in short, would war against Germany on land throughout East Asia, not just on the high seas. By November, Tsingtao, key to Shantung, had fallen to 30,000 Japanese troops who then began to spread throughout the province despite China's bitter protests. By that time, Japanese forces had also occupied Germany's Pacific islands.

Yuan Shih-k'ai still called himself President of the Chinese republic in the summer of 1914, but his hold on his country was tenuous. Even so, Katō at first worked with Yuan, because Yuan had the authority to declare Shantung a war zone, which would permit Japanese troops to operate there legally. As well, more cautious Japanese leaders, especially the *genrō*, feared that Yuan's downfall would mean anarchy, and anarchy might foster the growth of radical nationalist movements that might endanger Japan's position in China over the long term.[4]

This partnership was destined to be short. Yuan's worst fears were confirmed when Japanese forces occupied all of Shantung, not just the German leasehold area around Kiaochow. So the Chinese President withdrew his war-zone declaration and insisted that Japanese forces evacuate most of Shantung.

Katō's response was the 'Twenty-one Demands' note, conveyed to Yuan in January 1915. The demands were in five groups. The first would confirm Japan's inheritance of Germany's rights over Shantung. The second would extend Tokyo's leaseholds over south Manchuria for 99 years and enlarge Japanese rights there. The third dealt with Japan's existing and enlarging investments in the Hanyehping mining complex in central China. The fourth would bind China not to alienate (lease or transfer to another power) any part of the Chinese coast. The fifth group, which Katō termed 'wishes', not demands as such, would have made all of China subject to Japanese 'advisers', military, economic, political, even religious. Acceptance of these 'wishes' would have made China a sort of Japanese colony.

To Japanese eyes, the Twenty-one Demands were essentially defensive, except for the sweeping fifth group. The first three proposed, at most, only enlarging or extending rights Japan already enjoyed. Even the fourth group suggested only that China do nothing to endanger existing Japanese rights. Group Five, of course, was another matter. For this reason, Katō chose not to make it public to the other powers.

Yuan had no such scruples. Desperately seeking foreign assistance, he leaked the demands to Britain and the United States (and anyone else who would listen). But his timing was unlucky. The British, allied with Japan and at war in Europe, were not inclined to risk a breach with Tokyo for Yuan's sake. The Americans had neither alliance nor war. But President Woodrow Wilson was distracted by Germany's declaration of a submarine blockade around the British Isles in early February. His Secretary of State, William Jennings Bryan, was acutely aware of legislation pending in two Western states that would radically restrict the ability of Japanese immigrants to own land there. He hardly wanted a confrontation with Tokyo that would play into the hands of American nativists. Russia, the other potentially interested power in Asia, was too dependent upon Japanese military aid, especially ammunition, even to protest.

So Yuan employed another weapon of the weak: he delayed. His repeated cries for help finally persuaded the American Government to object, albeit mildly and only to Group Five. Even more cannily, Yuan sent an appeal to the *genrō*, arguing that Katō's aggressiveness was undermining Sino-Japanese friendship and risking friction with the West.

The *genrō* fully agreed. Their irritation with Katō had been growing for some time, especially over his refusal to keep them informed of the day-to-day conduct of his diplomacy. By early May, they had persuaded the rest of the cabinet to overrule Katō and drop Group Five from the demands altogether. Katō was able to issue an ultimatum to China, which at last forced Yuan's hand on groups one to four. But it was a pyrrhic victory. America's President Wilson proclaimed that the United States would not recognize any part of the Sino-Japanese settlement that violated its rights, the Open Door or China's territorial integrity. In China, rising public opposition to Yuan's humiliating surrender promised further difficulties there for both Yuan and Japan's position in China.

Most importantly, in Japan the *genrō* moved to oust Katō from the cabinet. This was no simple task, since Katō controlled the *Dōshikai*, which was a vital part of the cabinet's power in the Diet. But at the end of June, an election bribery scandal involving the *Dōshikai* broke, forcing Katō out.[5] The cabinet kept the party's support by appointing Katō's friends to be the new Foreign Minister and Vice-Minister.

Ishii Kikujirō's diplomacy was not a radical departure from Katō's. But the new Foreign Minister was more interested in ending Japan's relative isolation from the powers that his predecessor's aggressiveness had created. To this end, Ishii quickly promised that Japan would seek no separate peace with Germany. This was no merely symbolic declaration. It committed Japan more than ever to the cause of Allied victory. It also ensured that Japan would have a role in the peace conference that would follow that victory. Ishii was not so certain of the wisdom of accepting a military alliance offered by Russia, but he abided by the wishes of the *genrō* (led by Yamagata) and signed such an agreement in July 1916.

While Ishii was much more interested in good relations with the leading powers than Katō had been, he shared Katō's expansive ambitions toward China. In November 1915, Yuan moved to enlist international assistance by proposing that China declare war on Germany. Then China would be a *de facto* ally of Britain, France, Russia, and Japan, making any further Japanese encroachments a matter of alliance, not just Asian, affairs. Ishii vetoed the attempt.

Ishii even supported the Imperial Army's moves to unseat Yuan entirely. Japanese military aid helped a Mongol force defeat Chinese troops in north China in early November, as dissident Chinese generals visited Japan to solicit support openly. They received it under the guidance of the Imperial Army's Vice Chief of Staff, Tanaka Gi'ichi. By early 1916, these generals had become 'warlords', declaring their provinces independent from Yuan's government in Peking. By springtime, Ishii was able to force Britain, France, and Russia to join Japan in blocking Yuan's ability to collect the crucial salt tax in China. Yuan made preparations to flee the country, but died in early June. Southern China slid into the chaos of local warlord regimes. The north was in a similar state, unified only in name under 'Prime Minister' Tuan Chi-jui, who understood that he had to cooperate with Japan or suffer Yuan's fate. Overall, Japan had done well in China. Its position in Manchuria was secure, it had obtained the valuable province of Shantung, and the way was open for further encroachments with the consent, or at least acquiescence, of several European powers.

But the *genrō* remained concerned. They knew that the war in Europe would end eventually. Looking to the long term, they wanted good relations with the winning powers, namely Britain, France, and Russia. The *genrō* especially wanted to avoid estranging the United States. They feared that additional activity in China would sour ties with all four countries. They had domestic worries, too. Conservatives to the last, the *genrō* much preferred that the bureaucrats hold the upper hand in governing Japan, not the politicians. In this way, they saw Katō, Ishii, and the *Dōshikai* as a threat to Japan's relations with the great powers and to the bureacrats' domination of the Japanese Government. So in October 1916, they removed the cabinet and brought in General Terauchi Masatake, Yamagata's and the army's man, to be Prime Minister.

Terauchi immediately moved to diminish Katō's influence at home and abroad. Japanese elections held in April 1917 resulted in a huge loss of seats in the Diet for the *Dōshikai*.[6] In his foreign policy, Terauchi swiftly moved to improve relations with the Western powers by halting all consideration of further gains at China's expense.

Some improvement had come already, even before Terauchi became premier. At the *genrō*'s urging, Japan had signed an alliance with Russia in July 1916. It was not entirely an innocent document. In exchange for the alliance and increased shipments of arms and ammunition, Russia agreed to recognize concessions for Japan in northern Manchuria.

Japan played the same game with the British, who were increasingly hard pressed at sea by the start of 1917. The Terauchi Cabinet offered Japanese naval patrols in African and Mediterranean waters – something Tokyo had not even considered earlier – if Britain agreed after the war to support Japan's claims to Shantung and the northern Pacific islands seized from Germany. The Foreign Minister next moved successfully to secure French and Russian support for these claims.

In other words, while Terauchi did not intend to seek further gains in China, since these might antagonize the West, he was not going to relinquish any existing Japanese position there. This was acceptable to Russia, Britain, and France. But the United States might have been harder to win over. The Americans took the view that China ought to be allowed to determine its destiny itself (as Wilson put it, China had the right of 'self-determination'). Even Japan's existing position in China conflicted with such a right. To make matters worse, there was the 'Zimmermann Telegram' affair.

In January 1917, the British, hoping to nudge America closer to the Allied war effort, released an indiscreet document penned by a careless German representative which offered a German alliance to Mexico if the

United States entered the war in Europe on Britain's side. The Mexicans were to request that Japan join in an attack on the United States and, in the resulting triumph, Japan could have China while Mexico reconquered the American Southwest. The whole idea was preposterous, but some Americans remembered the war scares over immigration of a decade earlier and took the plot seriously. Fortunately, Japan was able to point out that it had played no role in the entire matter and ought to be held blameless.

Even so, Terauchi would have had a difficult time persuading Washington to tolerate Japan's gains in China since the start of the First World War in 1914. Ironically, the course of that war in the spring of 1917 enabled Japan to succeed. The German Government decided to permit its submarines, 'U-boats', to sink any ship – even American – on sight. This risky decision brought the United States into the war by April. Although Japan and America had signed no treaty of alliance, they were now fighting a common enemy. As importantly, in Terauchi, Japan had a leader who wanted to re-establish good relations with Washington – and knew how.

The key was China and the United States' commitment to Chinese 'territorial integrity' and right of 'self-determination' that had led to difficulties with Japan during the Twenty-one Demands incident. Terauchi and his Foreign Minister, Motono Ichirō, understood that Japanese encroachments against China or even manipulation of China's instability would spoil any chance for better relations with the Americans. Accordingly, they decided that Japan would no longer aid the southern rebels, loosely organized under Nationalist Party (*Kuomintang*) leader Sun Yat-sen. Instead, Terauchi sent a personal associate[7] to extend loans and other support to the northern leader, Tuan Chi-jui, as Motono orchestrated an international effort to have Tuan recognized as the only legitimate ruler of China. Japan also relaxed its opposition to China – meaning Tuan – joining America in declaring war against Germany. This important move meant that China, too, would be present at any peace conference after the war, to Japan's discomfort.

The Americans could hardly protest against any of this. Nor could they object to the centrepiece of Motono's strategy: sending his predecessor, Ishii, on a special mission to the United States to improve relations. The very fact of Ishii's former rank impressed the Americans with Japan's sincerity in mending fences. And Japan had good reason to be sincere: the United States' entry into the war and consequent rearmament effort had led Washington to restrict the export of American steel. The Japanese shipping construction industry, a major beneficiary of the European conflict, had been hard hit.

Nevertheless, China remained the central issue in Ishii's discussions with American Secretary of State, Robert Lansing. And Ishii, an old associate of Katō, was not about to surrender any of Japan's recent gains there. Ishii opened his conversations with Lansing by stressing Japan's special relationship with China. He argued that the United States should recognize that relationship, as Britain, Russia, and France already had.

Lansing did not entirely agree. While he was not inclined actively to contest Japan's interests in China, he did hold those interests – especially the ones gained during the war – as contrary to the principles of the 'Open Door' and therefore damaging to China's long-term stability and unity. President Wilson was moving toward the more radical position that all powers' interests in China, no matter how old and established, ought to be liquidated. Fortunately for Japan, Wilson wanted no confrontation with Tokyo while the United States was at war with Germany. But the resulting 'Lansing–Ishii notes' of early November were disappointing anyway. The Americans allowed that 'territorial propinquity creates special relations', but they would not agree to recognizing Japan's 'interests' anywhere in China. This did not bode well for any peace conference once the European conflict was over.

Even so, the Terauchi Cabinet had reason to feel satisfied. The Prime Minister and Motono largely had been able to end Japan's diplomatic isolation, and in such a way that Britain, France, and Russia had recognized Japan's wartime gains in China. There, Japan's new policy of aiding Tuan inspired hope that he would prove to be powerful enough to subdue radical Chinese nationalism and pliant enough to submit to further Japanese influence.

But all of Terauchi and Motono's finely tuned calculations were thrown awry just as the Lansing–Ishii notes were exchanged. The Russian monarchy, which had been on increasingly soft ground after the Russo-Japanese War, collapsed completely under the pressures of the far larger First World War. By early 1917, it had succumbed to a makeshift regime of the departed Tsar's ministers and supporters and moderate opposition elements. None of these groups had a firm grip on power; they fell to the Bolsheviks in October 1917 as the Russian Empire slid into a deep and bitter civil war.

The fall of Russia was a terrible blow to Japan's diplomacy. The most immediate effect was to make the recently signed alliance with the Tsar worthless. The new Bolshevik regime was not bound to recognize Japan's new position in China and, especially, Manchuria. It became apparent, just as immediately, that the Bolsheviks indeed did not consider themselves so bound. In fact, they made public all of the old Tsarist regime's secret treaties with Britain, France, and Japan and denounced them.

These setbacks, important though they were, seemed insignificant in light of the greater potential for disaster posed by the Russian Revolution. The *genrō* and the bureaucrats were unwavering conservatives. They feared that the revolution would renew calls within Japan for radical reforms such as workers' rights and suffrage for all adult males. In the years before the First World War, the Japanese Government had strongly opposed these reforms, even executing anarchists and socialist leaders. The conservatives openly condemned ideas of 'natural rights'. They were predictably horrified by the Bolsheviks, who formed workers' soviets and who routinely did away with the aristocracy and bureaucrats of the old regime.

Frankly, the conservatives' grip on power was much too firm in Japan to permit much progress for any reform movement there in the decade beginning in 1910. But Japan was unable to contain the impact of the new ideology of communism upon the turbulent situation in China. In southern China, the *Kuomintang*'s leaders had shown some willingness to embrace Bolshevik-style reforms. If they overcame Tuan, perhaps even with Bolshevik support, the nightmare of a virulently anti-Japanese, communist China allied with a Red Russia might come true.

Yet another catastrophe seemed even more imminent. Upon assuming power, the Bolshevik leader, V.I. Lenin, swiftly moved to take Russia out of the war and sign peace terms with Germany. Those terms ensured not only that the Germans could swing their vast forces in Russia to the West, against France and Britain, but also that the new Soviet Union would actually provide food and fuel for the Germans to continue their war. Terauchi and Motono, whose diplomacy had carefully built an allied consensus supporting Japan's gains – at Germany's expense – in Asia and the Pacific, now confronted a shattered consensus and a torn, perhaps defeated, alliance.

Nor were these the end of Japan's woes. Every Western leader groped for a way to bring 'Russia' back into the war. Every way posed problems for Japan. The French and British suggested re-establishing a front in Russia with French and British and (mostly) Japanese troops. When Terauchi and his Imperial Army dismissed this fantasy out of hand, Paris and London pressed for, at a minimum, a Japanese occupation of Asian Russia: Siberia. Terauchi was hardly keen on this idea, either. In January 1918 he was privately briefing other Japanese leaders that the European war was likely to last well into 1919 and Germany was likely to win it. Given this judgement, the Japanese Prime Minister, and the *genrō*, thought a Siberian expedition unwise. It would only alienate the Soviets who appeared, through their acceptance of those harsh peace terms, to be virtual junior allies with Germany. Why invade Siberia, in short, for a losing cause?

Even without the Siberian expedition, in a postwar world with Germany and, indirectly, the Soviet Union victorious, Japan's position would be in grave danger unless Tokyo had its own allies. Here, the likeliest candidate was the United States, certain to be a major, perhaps the major, world power after the war regardless of which European side prevailed. Yet President Wilson's proposal for keeping the Soviet Union in the war was his famous 'Fourteen Points' address of January 1918. Wilson's speech was carefully crafted to appeal to the Soviets. The American President called for 'open covenants, openly arrived at', echoing Lenin's denunciation of the secret treaties. Wilson called for Germany's evacuation of territories conquered against the will of their peoples. He specifically denounced the stern terms Germany had forced upon the Soviet Union. The right of each nation to determine its own destiny, the American President declared, was sacred to God's order and the future peace of humanity on earth.

Although delivered for Russian and German audiences, Wilson's words made for uncomfortable listening in Japan. In one sweep, Wilson appeared to be backtracking on the Lansing–Ishii notes' acquiescence to Japan's position in Manchuria, challenging outright Japan's wartime gains in China, and virtually offering alliance to Chinese nationalists under the banner of self-determination.

In response, Terauchi adopted a pragmatic approach. Ignoring Wilson's grand rhetoric for the moment, he replied to the Anglo-French proposal for a Siberian intervention by insisting that America's approval was a necessary precondition for Japanese action. This careful posture would ensure that the United States (not to mention the Soviet Union) would not be alienated by any unilateral Japanese action.

This approach was cautious and sound toward the great powers. But it became increasingly unpopular inside Japan as fears grew over the communist menace. Motono, who had served as Japan's ambassador to Tsarist Russia for many years, was appalled by the new Soviet Union. He argued for an immediate invasion of Siberia to keep it out of the communists' hands. Motono had a powerful ally in the Imperial Army. Its senior officers continued to believe that the restoration of Russia, or even a Russian front, was impossible. The best way to deal with the new Bolshevik regime was to prevent it from gaining influence in Asia by creating a pro-Japanese regime in Russian Siberia and preventing revolutionary elements from coming to power in China.

By the end of 1917, moreover, the Imperial Army was increasingly a force unto itself in the execution and conduct of Japanese foreign relations. During the early Meiji period, years before, the army had never been absent from foreign-policy making, but it had been restrained. Its leaders had never been in a position to strongly object to the government's foreign policies because its leaders, such as Yamagata, were vital parts of the government, and a government that acted through consensus achieved by the *genrō*. It never had used force on its own. It never had submitted substantial legislative proposals to the Diet beyond, of course, its own budget, and even this went through the cabinet first. The army always had yielded control of the fruits of its conquests to the government. Taiwan and Korea, won after the Sino-Japanese War of 1894–5, went to civilian control. After the Russo-Japanese War ten years later, a public (civilian) corporation, the South Manchurian Railway, administered Japan's sphere of influence in that part of China, not the army.

But by 1917 the army's position was changing. After Itō's assassination in Korea in 1909, the Governor-General of Korea typically was an army general. The Imperial Army controlled Shantung after it was taken from Germany in 1915, not a civilian agency. Army officers aided friendly local Chinese leaders and warlords, sponsoring risings and rebellions in Shanghai, Yunnan, and, most importantly, Manchuria and inner Mongolia (which the Japanese referred to as 'Man-Mo'). By the time of the Russian Revolution, the Imperial Army had its own officers, its own agents, and its own influence throughout much of China, especially those areas closest to the new Soviet Union.

When the French and British request for intervention in Siberia came in early 1918, therefore, the Imperial Army moved to use such an intervention for its own purposes. It moved to locate friendly Russian leaders and warlords-to-be. In essence, the army sought to create pro-Japanese local regimes in nearby Russian territory in much the same way that it had done in China. At first, it seemed to enjoy similar success. The Russian administrator of the Chinese Eastern Railway in northern Manchuria, appointed by the ousted Tsar, preferred to collaborate with the Imperial Army rather than risk himself at the mercies of the communists. Cossack leader Gregorii Semenov saw an opportunity to create, with Japanese help, a new nation for his people in eastern Siberia, with himself conveniently at its head. And ex-admiral of the Tsar's navy, Alexander Kolchak, arrived there to offer his services to any anti-communist Russians – or foreign governments – that might wish to support him.

Siberia was not China, however. The Chinese Revolution of 1911 had ended the authority of the Manchu Emperors without putting anything in its place. The Chinese Empire simply fell apart. The Bolshevik Revolution of 1917 was spearheaded by a dynamic force, the communists, who had the supreme advantages of organization, dedication, and the support of a powerful central regime in the European parts of the Soviet Union. Throughout the winter of 1917–18, the Red Army drove east, capturing town after town along the strategic Trans-Siberian

Railway and moving closer each day to Manchuria and Japan.

Precisely because the communists were powerful, many Japanese leaders understood that an occupation of Siberia could turn into a very large and very long conflict. With the stakes so high, some of these leaders opposed an expedition to Siberia. To Terauchi's dismay, *Seiyūkai* leader Hara led the opposition, meaning that the majority of the Diet also would dissent. Even the *genrō* were split on the question.[8]

The Imperial Army was appalled at the government's indecision and delay. By the end of February 1918, it could wait no more. Vice Chief of Staff, General Tanaka Gi'ichi, coordinated two efforts. One saw military aid – weapons, ammunition, and other supplies – flow to Semenov. The second accelerated planning for a large, direct Japanese intervention into northern Manchuria and Russia itself, to seize the Trans-Siberian Railway all the way to Lake Baikal.

Naturally, the army did not want to fight Chinese troops in northern Manchuria.[9] That sort of conflict would bring Western criticism. It would also undermine Terauchi's policy of cultivating friendly Chinese leaders such as Tuan. So, in March, the Imperial Army (again acting under its own authority) reached agreement with Tuan's regime – as the official government of China – giving Japanese troops the right to engage in 'joint' operations with Chinese forces around the railroad to combat communism.

The Imperial Army hoped to reforge a consensus among the Japanese élites with its initiatives in Manchuria and Siberia. Instead, it created deeper divisions. Motono, who had favoured intervention in Siberia, was outraged that the army had conducted its own diplomacy with China completely beyond his Foreign Ministry's control. Not only was this a slight to him and his ministry, but it also complicated Japan's position with the United States. Motono saw America as the key to his strategy. If Washington joined London and Paris in requesting that Japan undertake a Siberian expedition, Japan could hold on to any gains realized there with the open consent of the international community. The army's independent acts threatened all this. Its unilateral steps already had raised suspicions in Washington about Japan's ultimate objectives in Siberia, Manchuria, and China, too. Some American officials, who had never liked the Lansing–Ishii compromise, began to argue that much stronger measures were needed to contain Japanese ambitions.

Hara and those *genrō* opposed to a Siberian expedition saw matters differently. If the Americans really were suspicious of Japan, why not simply avoid intervention altogether? In their eyes, Motono was something of a schemer. He was angry that the army had acted without him, but still sympathized with what the army had done. When Motono dropped his objections to the military's unilateral decision to use force in

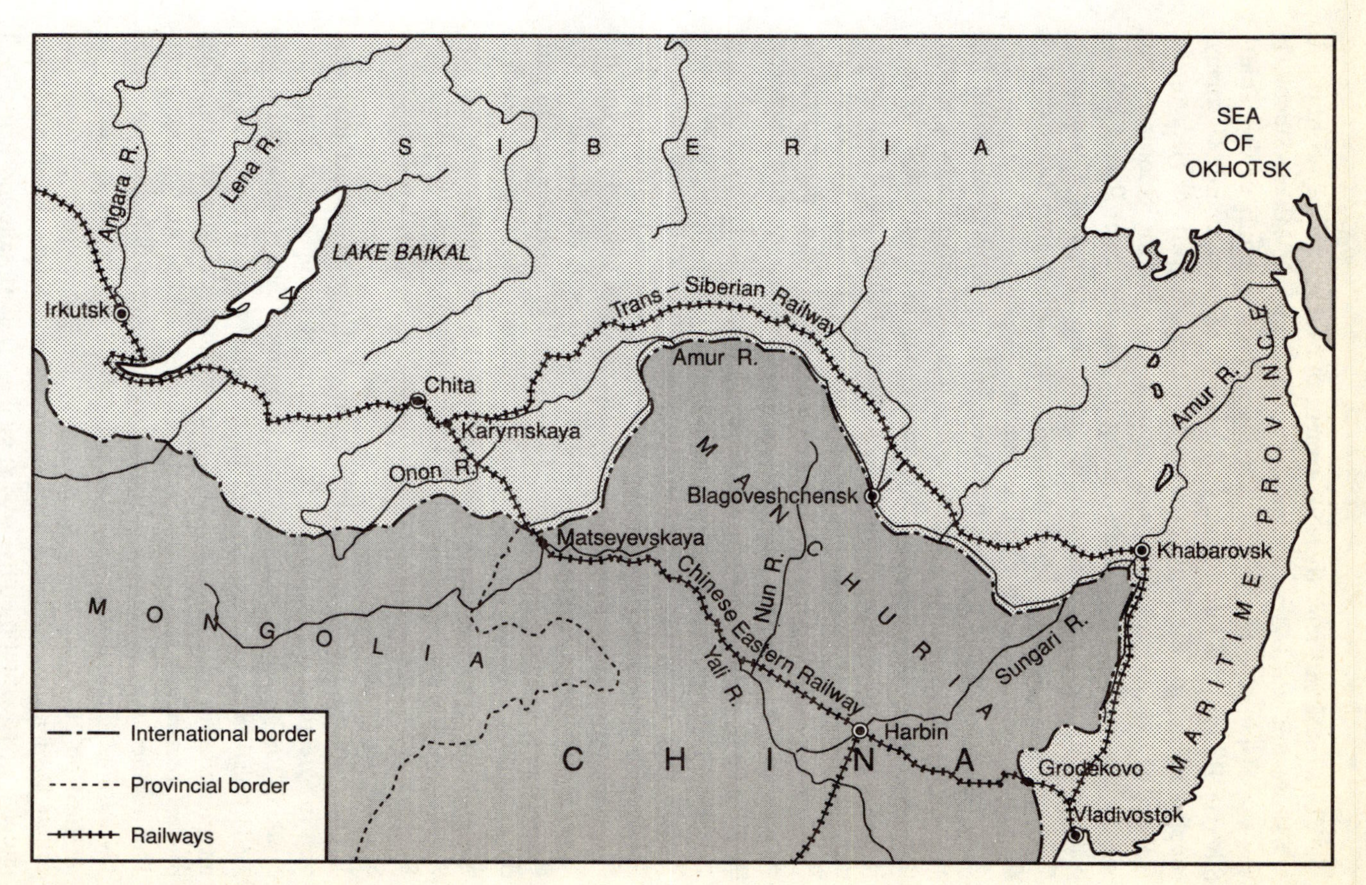

Map 4 Eastern Siberia 1918

the Vladivostok incident, their judgement of Motono seemed confirmed.

Vladivostok was by far the most important city in eastern Siberia. It was the major port of the area and the eastern terminus of the Trans-Siberian Railway. While Russia had been in the war against Germany, supplies meant for the Russian army had been shipped to Vladivostok. Control of the city, therefore, was important not just for its location, but also for the war materials stacked along its wharves. After Russia left the war, Britain, the United States, and Japan had each sent a warship to Vladivostok to prevent those materials from falling into the hands of Germany, German agents or plain bandits.

At first, no Japanese marines had been landed in Vladivostok itself. But as the communists marched east along the Trans-Siberian Railway toward the city, pro-communist disorders broke out in Vladivostok. After a Japanese was killed on 4 April, the Japanese warship landed marines, as did the British (but not the American). The landing backfired. All Russians, communist or not, were outraged. Anti-foreign demonstrations erupted all over the city. The warship commander requested reinforcements, precipitating a crisis in Tokyo.

Motono argued his case for intervention in all Siberia, not only Vladivostok. Germany, he maintained, was using the communists as its dupes. And the communists threatened Japan's position in China, if they could not be stopped in Siberia. But Motono was alone. Hara and the *Seiyūkai* were openly dubious. Terauchi was ill with diabetes. Even the Imperial Army refused to support Motono's call for intervention. In part, the army was irritated with Motono's complaints about its independent diplomacy. As importantly, however, the army's leadership was angry that it had been the navy that had first landed troops on Russian soil. The army wanted a Siberian intervention, but only so long as it was the army's Siberian intervention.

The resulting paralysis in Tokyo required the *genrō*'s direct intervention in foreign-policy making – the last time it would occur. Yamagata, speaking for his ailing protégé Terauchi, insisted that Japan avoid diplomatic isolation at all costs. To escalate the Vladivostok crisis risked direct intervention in the Russian civil war without the support of other powers, particularly the United States. There would be no reinforcement. The marines were to leave Vladivostok as soon as possible. And the chief advocate of immediate escalation, Motono, was to leave office. He resigned on 10 April.

The Vladivostok crisis had been contained and Motono dismissed. But the root of Japan's difficulties was as vital as ever. Britain and France still wanted a Siberian intervention. The United States still held back. The communists were consolidating their control over all Siberia – Vladivostok fell to them by early May. And Germany was about to open an offensive in Europe that could well win the war. How was Japan to meet these challenges, both near-term and far-term?

The new Foreign Minister, Gotō Shimpei,[10] believed that Japan had to take the long view. In his judgement, Britain and France were exhausted. Whether they won the war or not, they would no longer play a major role in global, especially Asian, politics. The future belonged to Germany and the United States. Germany was preying upon Russia, using the communists as its pawns. In some respects, however, that was reassuring. It was true that Russia had taken itself out of the European war. It was also true that Germany itself had elected not to play an active role in Asia. In Gotō's calculus, then, Britain and France were prostrate, Russia and China were in pieces. Only the United States and Japan remained real powers, real players in the game, in East Asia.

This judgement naturally led Gotō to evaluate the Americans' intentions, and he found them to be aggressive. The immediate cause of Gotō's concern centred on an American railroad commission that had arrived in Vladivostok shortly after the United States had entered the war in Europe. The commission, made up of civilian railroad experts, had been devoted ostensibly to speeding up the transport of supplies from Vladivostok's port to (then Tsarist) Russian troops fighting Germany. Gotō suspected that the Americans secretly wanted to purchase all of the Russian railroads – the Chinese Eastern and parts of the Trans-Siberian – in a bid for American control of northern China and Siberia. Nor was Gotō alone in these suspicions. He brought with him into the Foreign Ministry an impressive array of hardliners: Arita Hachirō, Hirota Kōki, and Matsuoka Yōsuke, all of whom would make their mark in the decades to come.

Their immediate mark, though, was to be made in early 1918 in Siberia. Gotō and his lieutenants strongly favoured intervention. Like the departed Motono, they wanted Allied and American approval first. They received help from an unlikely source: a group of Czechs.

The Czechs were a discontented nationality within the Austro-Hungarian Empire. When war had broken out in Europe, a number of them crossed lines to fight alongside the Russians against their Austro-Hungarian masters. By early 1918, there were nearly 50,000 men in the 'Czech Legion'. Stranded by Russia leaving the war, they resolved to make their way to Western Europe by the only route left open: the Trans-Siberian Railway to Vladivostok and thence by ship to continue their fight from Italy and elsewhere. The Czechs reached agreement with the communist leadership for their safe passage by train to Vladivostok in exchange for their disarming themselves. Neither side fully honoured these terms. As a result, the accord broke down in late May with about a third of the Czechs safe in Vladivostok, while the rest, strung out along the railroad throughout Siberia, began to seize key stations and cities.

Japan cared little for the fate of the Czechs in Europe. But Woodrow Wilson was another story. Here was a clear case of a people struggling for their right to determine their own fate. The plight of the Czechs moved Wilson to approve American intervention in a multinational Siberian expedition to assist the Czechs' evacuation.

The Czechs' seizure of key points in Siberia and America's green light for intervention revolutionized the disposition of Siberia and ended the political impasse within Japan. As the Czechs seized each city in turn, anti-communist Russians materialized to declare themselves the rightful rulers of their corner of Siberia. Admiral Kolchak stepped forward to declare himself the best ruler of these rulers for all Siberia, and quickly received British and French support. The Czechs, interested in leaving, not ruling, Siberia, moved to occupy the Chinese Eastern Railway in northern Manchuria as the most expeditious way out.

All of these developments threatened Japan's position. Semenov, their Cossack leader, appeared increasingly irrelevant and useless. Kolchak was Britain's and France's man. And a Czech occupation of the Chinese Eastern Railway would overthrow the position Japan had so recently won there in collaboration with Tuan's regime in China.

Much of this difficulty could be removed, of course, by a direct application of Japanese military might. But there were two restrictions. One was America's objection to anything more than rescuing the Czechs. Intervening forces were not to seize control of railroads themselves, nor were they to support anti-communist regimes in Siberia. To ensure no mischief, the expedition would have no (Japanese) supreme commander. Its various national forces would operate on a 'joint equal' basis.

The second restriction came from Hara. He was willing enough to approve the kind of limited expedition that the Americans stipulated. Vladivostok could be patrolled, and arrangements to avoid misunderstanding along the Chinese Eastern Railway seemed prudent. But Hara suspected that the Imperial Army had far more in mind.

Hara was correct. The army had drawn up plans for nearly 50,000 troops to proceed along the Trans-Siberian Railway to Chita, almost all the way to Lake Baikal to create an anti-communist, pro-Japanese regime over all eastern Siberia, backed by Japanese military assistance and, if necessary, firepower.

In the showdown on 16 June, Hara insisted upon a small expedition. It would secure order in Vladivostok and protect – but not control – the railroads around that city. No more than two divisions (that is, far fewer than 50,000 troops) were to be sent unless the government – not the army on its own – explicitly authorized an increase.

The Americans knew nothing of these internal divisions. Given Japan's recent moves against China, however, they were wary even of a general dispatch of Japanese forces to the area around Vladivostok, rightly suspecting that this might well include the Chinese Eastern Railway. Convinced that such a broad sweep would impinge on the self-determination principle of the Open Door, Washington rejected Japan's proposed terms for intervention in late July and insisted that Japan and the United States send not more than one division each, though the Americans did change their stance to permit and offer the Japanese the honour of overall command. Terauchi and Gotō were angry, the Imperial Army furious, at the Americans' inflexibility.

But it was intervention under Hara's and America's terms or no intervention at all. So the army agreed to limit its force in the expedition to a single division to Vladivostok, so long as the Japanese Government understood that a second might be sent, if needed, to positions beyond Vladivostok. Under these rather rigid stipulations, Japanese troops began landing in Siberia in early August.

The Imperial Army knew that once Japanese troops had landed in Siberia, it would dominate Japanese policy towards Siberia regardless of Hara's objections. It may go too far to say that the army's leaders consciously intended to deceive Hara and the Diet, but many in the army, and in the government, too, believed that politicians could not be trusted to run Japanese foreign policy. In this case, the army had no intention of allowing existing gains in Manchuria to be jeopardized by Hara or the Americans. Nor was it about to pass up the opportunity to create a friendly, non-communist regime in Siberia. By October 1918, nearly four divisions had been sent to occupy most of the Trans-Siberian Railway east of Lake Baikal. By the end of the year, four more were in Siberia, making a mockery of the conditions Hara had set on their dispatch. This was no small matter because, even before October, Hara Kei had become Prime Minister of Japan.

Ironically, it was not the Siberian Expedition that forced ex-general Terauchi from power, it was rice. More broadly, that is, the Terauchi Government's inability to cope with rising rice prices and other economic disturbances caused by Japan's wartime boom brought it down, opening the way for a substantial realignment of its foreign relations.

The Japanese economy enjoyed a tremendous boom, spurred by the war in Europe. The number of factory workers, for example, doubled to over two million between 1914 and 1918. The boom created inflationary pressures raising the price of many basic commodities. Incomes rose too, but not as quickly and not evenly throughout the country. The heads of the *zaibatsu* combines did spectacularly well. The boom swelled their purses and enabled them to acquire many smaller companies which lacked equal financial resources, and to start up new branches of their own.

Mitsubishi, to name one such *zaibatsu*, created nine new Mitsubishi companies toward the end of the First World War, including the huge Mitsubishi Shipbuilding Company, Mitsubishi Internal Combustion Engine, and Mitsubishi Steel. Mitsui went into the chemical and synthetic fabric businesses.

The *zaibatsu* increasingly used their financial clout to influence (and often, simply buy) members of the Diet. Workers in Japan's new industrial sector, which overwhelmingly meant people who worked for one or another of the *zaibatsu*, also prospered, though their lives were by no means comfortable. But, even by 1918 and the end of the wartime boom, most Japanese were not industrial workers. Over half remained in the traditional sectors of fishing, mining, and farming. The result was increasing difficulty for the great majority of ordinary Japanese. That difficulty turned to disaster when the price of rice – the fundamental basis of the Japanese diet – soared in the summer of 1918. Demonstrations against the high rice prices turned into riots directed against rice shop owners and, inevitably, the government. Beginning in agricultural areas, the protest spread to urban workers and, most militantly, to miners. By mid-September, over half a million Japanese had joined the riots. Terauchi felt he had no choice but to use the Imperial Army to restore order, at times with fixed bayonets. Over a thousand people were killed or wounded in the process.

The rice riots of 1918 finished Terauchi. With blood on his hands, he had to resign. Who would succeed him? It was a reflection of how little democracy there was in Japan that no political leader could claim close ties to the rioters, or even their concerns. Katō had played demagogue, openly courting the rioters with promises to obtain the right to vote for all adult Japanese males. He did not do so from any genuine feeling for the masses, but rather because he realized that he had become *non grata* to Japan's existing political order. But though the rice riots had driven out Terauchi, they had done nothing to touch the highly conservative political order itself. The original *genrō* had nearly all died, but the survivors[11] spoke for the civilian bureaucrats, the military, and the business community in vetoing Katō as the next Prime Minister. That left Hara, who craftily sealed his appointment as premier by naming Terauchi's (and hence Yamagata's) protégé, Tanaka Gi'ichi, to be the new Army Minister.

No sooner had Hara come to power than the great war in Europe unexpectedly ended in November, and Japan, as with all the other powers, was swept up into the peace conference at Paris that would decide the shape of the world in the years to come.

Japan's aims at Paris had been established long before Hara had become Prime Minister. The territorial gains of the northern Pacific islands made at the start of the war had to be confirmed. Rights and concessions won in Shantung at China's expense were to be acknowledged. The current business in Siberia was to be kept off the table entirely. In other words, Japan sought to have its enlarged imperial position recognized by the international community.

The chief obstacle to all of these objectives was the United States, particularly President Wilson. Wilson had arrived in Paris with great fanfare as the bringer of a new world order. The keystone of this new order would be a League of Nations, where all nations would be represented as equals and none allowed to be the victim of another's aggression. Wilson meant to revolutionize the rules of the game of international relations in ways nearly always detrimental to Japan's hard-won gains over the preceeding quarter century.

Japan's actions in Siberia, Shantung, and the North Pacific were hardly in keeping with these new rules. As worrisome, Japan's prewar position in Manchuria might not be allowable either. A further complication arose from a vigorous press campaign in Japan to take the League of Nations at its word and insist upon a clause in its charter guaranteeing the principle of racial equality among nations. As Hara knew, ideologically (and logically) the proposal was unimpeachable. But to insist upon it might revive American – and, increasingly, Canadian and Australian – sensitivities over their immigration policies.

In addition to dealing with these problems of policy, Hara also had to contend with the politics of choosing a delegation for the peace conference. His own Foreign Minister was ill. Katō had extensive diplomatic experience, but as leader of the opposition *Kenseikai* his appointment was out of the question. Ishii was passed over, too, because of his prior ties with Katō. Instead, Hara secured the appointment of two political reliables: Saionji Kimmochi and Makino Nobuaki.

Both men were popular with the public. This was vital to Hara and to the government in general. Memories of the Japanese public's bitter rage against the disappointing peace terms ending the Russo-Japanese War in 1905 were still fresh thirteen years later. Hara, who had used the public's anger to catapult his *Seiyūkai* into power, wanted to hand his opposition no similar opportunities in 1918.

For both diplomatic and political reasons, therefore, Japan approached the Paris Peace Conference defensively. It sought to avoid being deprived of what it already had, and Hara hoped to avoid a popular protest in case some deprivation occurred anyway.

Japan's strategy turned out to be entirely appropriate. The Paris conference opened in January 1919. Although Japan was widely recognized as one of the great powers, only the 'Big Three' – Britain, France, and the United States – determined the most serious issues. Although Japan had obtained Chinese recognition of Japanese rights in Shantung (and

Manchuria) in prior agreements, Chinese delegates at Paris bitterly attacked Japan as a violator of the spirit of the League of Nations. Although Japan physically occupied the former German Pacific islands north of the equator, Australia and New Zealand, using their influence on London as British Dominion countries, objected to Japan's presence there. These countries and others concerned about non-white immigrants, most importantly America, also blocked Japan's racial equality clause from being inserted into the new League's Covenant.

Alarmed that its delegation might return from Paris with nothing to show for its efforts, the Japanese Government moved to refuse to join the League of Nations unless Japan's minimum position in Shantung (gaining German leasehold rights around Kiaochow, but relinquishing them elsewhere in the province) were granted. Saionji and Makino privately but vigorously objected. They argued that the League represented the new order, at least to the war's victors. To stand aloof would put Japan in the company of the international community's pariahs, Germany and the Soviet Union. Instead, they offered to surrender any German right that infringed upon China's sovereignty. This concession, coupled with rumors of the Japanese Government's threat, led the 'Big Three' to agree to the transfer of German rights to Japan. While the agreement ensured that Japan would join the League of Nations, it led to bitter disappointment in China and a massive outpouring of nationalist outrage there when revealed in May 1919.

Japanese concessions paved the way to settlements of the other issues as well. Japan could retain the northern Pacific islands, but not as sovereign possessions. Instead, they would be administered by Japan as 'mandates' for the League of Nations and, as such, could not be fortified. As for the principle of racial equality, Makino vowed that his country would strive to have it recognized by the new organization as soon as possible.

In truth, Hara had a far broader agenda for Japan. He was convinced that the key focus of Japan's foreign relations in the years ahead would be the United States. To ensure cooperation with America, and the principles of the 'new order' that America favoured, Hara had willingly paid a price at Paris to begin Japan's membership in that order. He would continue to pay over the next two years to secure that membership.

There were three chief issues of contention between Tokyo and Washington: Japan's enlarged Siberian Expedition; a growing naval race among the leading seapowers of Japan, America, and Britain; and Japan's newly confirmed rights in China. Hara was anxious to resolve them all.

He moved quickly in China, on two fronts. At America's insistence, he recognized the right of British, French, American, and Japanese banks to form an international banking consortium to coordinate lending to

China for economic development – despite the Imperial Army's opposition, especially to lending for projects in Manchuria or inner Mongolia. Japan formally agreed to join the consortium in May 1920.

As well, Hara authorized direct negotiations with China to oversee the withdrawal of Japanese forces from Shantung. He was disappointed in this regard. The Chinese, who had refused to agree to the Paris conference's settlement of Japanese rights in Shantung, also refused to do so in any bilateral discussions in Tokyo. They were hopeful that a Pacific conference of interested powers would be held soon. There, they could count on a friendly United States to help them secure an arrangement much better than any they could negotiate with Tokyo directly.

Hara, too, anticipated the possibility of a Pacific conference, which gave him all the more incentive to clear the air with America as rapidly as possible, especially over the growing quagmire in Siberia. By the start of 1920, it was obvious that the communist regime in Moscow would successfully establish control over Siberia unless the great powers wanted to resist militarily. The United States had no desire for war with the communists and began withdrawing its remaining forces from Siberia as rapidly as possible.[12]

Hara came to a similar conclusion. He wanted the Imperial Army out of Russian territory, too. But the army disagreed. Normally, Hara as Prime Minister would have had little influence with General Uehara. But Hara had wisely chosen General Tanaka Gi'ichi as his Army Minister. Tanaka was an ambitious general who hoped for the prime ministership one day. He retained influence within the army, but he also established an alliance with the *Seiyūkai* by cultivating Hara. As a result, Tanaka coaxed the army into withdrawing from the Lake Baikal region as a first step to leaving all Soviet territory. But in March, some Russians killed a group of Japanese in a fishing settlement at Nikolaevsk, giving army hardliners a perfect excuse to delay their pullout. It took further intervention by Yamagata, the last of the original *genrō*, to secure Hara's ends. Even then, the army insisted on occupying the northern half of Sakhalin island until the Soviet Union apologized for the murders.

Hara was interested in ending the Siberian Expedition not only to improve relations with the West. He had come to power in September 1918 as a result of the deteriorating Japanese economy that had given rise to the rice riots. The chief culprit was inflation and the traditional cure for that was to reduce government spending. Since the military services took up the lion's share of the government's budget, they would have to bear the brunt of any spending cuts to be made. Fortunately for Hara, he and his *Seiyūkai* represented the lesser of two evils for the army and navy, as Katō's *Kenseikai* had gone radical, calling openly for universal male suffrage and extreme reductions in the military's budgets.

Hara had Tanaka's influence to keep the army under control. The Imperial Navy was another matter. During the war in Europe, it had carefully kept pace with the army in insisting upon ever more spending for ever larger forces. But the army's unauthorized escalation of the Siberian Expedition had disrupted this army–navy balance and angered the admirals. The navy, however, knew that Hara was displeased with the enlarged expedition. So in October 1918, shortly after Hara became Prime Minister, his Navy Minister Katō Tomosaburō (not related to politician Katō Takaaki) agreed to defer further increases in the navy's construction budget so long as Hara would protect the navy's existing programme to achieve the '8–8 Fleet,' that is, a fleet formed around eight battleships and eight battlecruisers. Hara and his *Seiyūkai* agreed to this proposal and delivered the budget by the summer of 1920. Through his cooperation with Tanaka and Admiral Katō, Hara had achieved the near-miracle of tranquillity over military spending.

The hard-won consensus ended within a year. In the spring of 1921, the United States invited all interested powers to a conference in Washington, DC, to discuss Pacific issues and naval arms limitations. Hara was well prepared to deal with Pacific issues, but to raise naval ones threatened to undermine his agreements at home with both navy and army. To refuse to come to Washington, though, would neatly torpedo his overall objective of good relations with the Americans. It was a most uncomfortable dilemma.

To Hara's relief, Admiral Katō argued strongly for going to Washington and accepting limits on naval construction. The Navy Minister pointed out that the United States was in the midst of a colossal building programme, one Japan could not match. Was it not better to see the Americans restrained so long as a respectable Imperial Navy were maintained? To ensure that respectability, Admiral Katō himself volunteered to journey to Washington. To ensure that no hardline rival supplanted Katō while he was out of the country, Hara agreed to serve as interim Navy Minister himself.

All of these developments troubled the Imperial Army. It was fine if the navy wanted to accept budget cuts itself, but the army suspected that the same principle of army–navy balance that had acted to bolster both service's budgets in the past would now be used to demand that the army follow the navy down the path of retrenchment. In addition, for a civilian – a leader of a political party, no less – to become a service minister horrified the army, which consented to the arrangement only after gaining explicit assurances that it would never be applied to the army. Not least, there was the matter of Japan's positions in Asia, positions usually held by the Imperial Army. The Americans were not going to leave these unchallenged. The army suspected the navy would be only too eager to strike a respectable deal over naval limits at the expense of the army's continental interests.

Allaying these army fears was another test of Hara's political acumen, and another test that he passed. Hara ensured that the Washington Conference would not discuss 'bilateral issues', meaning those solely between Japan and China, and Japan and the Soviet Union. That is, neither Shantung nor Siberia would be subject to Western interference. As well, Hara agreed to defer pressing for the army's swift withdrawal from these areas. He further promised that Japan would agree to nothing sacrificing its far-reaching rights in Manchuria. These assurances were good enough for the army and so, in August 1921, Japan officially agreed to attend the conference.

The Washington Conference opened on 12 November 1921 with an electrifying speech by American Secretary of State Charles Evans Hughes. The American proposed not just limits on naval armaments but actual reductions. Managing a quick recovery from his initial shock, Admiral Katō indicated Japan's basic agreement, so long as the navy's newest battleship would not have to be among those scrapped. He even went along with the American proposal that the battleships of the three largest navies, British, American, and Japanese, be limited according to a ratio of their tonnage: 5 tons of British battleship to 5 of American to 3 of Japanese. In exchange, Admiral Katō insisted that the two Western powers agree not to fortify their Pacific possessions (except Singapore for Great Britain and Pearl Harbor for the United States). The West agreed.

But the Tokyo Government almost did not. Hara Kei's brilliant balancing act of politics and diplomacy, his brilliant career, came to an abrupt and tragic end on 4 November 1921 when he was assassinated by a young fanatic. No one could fill Hara's shoes. Takahashi Korekiyo became Prime Minister and *Seiyūkai* leader. He was a superb financier but an incompetent politician. Foreign Minister Uchida was an experienced diplomat, but he had no political base. The consensus for Japanese foreign relations that the *genrō* had maintained through their stature, and that Hara had maintained through his political skill, now threatened to break down with consequences, both foreign and domestic, that no one could predict. But one thing was certain: Hara's assassination was not a happy way to conclude Japan's still-unfinished business at Washington, or to begin Japan's uncertain participation in the 'new order'.

Notes

1 In essence, each 'combine' or conglomerate was made up of a central holding company, itself usually centred around a bank, that owned controlling interest in a wide variety of other companies. For example,

there might be a Mitsui mining company, and a Mitsui shiping company, all owned by the central Mitsui holding company centred around the Mitsui bank.

2 Katsura was Yamagata's protégé, just as Saionji had benefited from Itō's sponsorship. In this way, the old rivalry between Yamagata and Itō was passed to a new generation.

3 The explanation for Saionji's decision is simple. Most of his personal supporters came from the old province of Satsuma. Since the Meiji Restoration, Satsuma had supplied most of the leaders of the Imperial Navy (just as Chōshū men dominated the Imperial Army).

4 The *genrō* were right. Radical nationalist movements already were beginning to emerge in China, such as the *Kuomintang*, a small group operating in the southern Chinese city of Canton. Its leader, Sun Yat-sen, had studied in Japan and had begun his revolutionary activities to oust the Manchus from Japan. He had a wide array of Japanese supporters, from radical nationalists such as Kita Ikki to radical Japanese socialists.

5 Bribery was an accepted part of every election, of course. This scandal, as with most others, became an occasion of public disgrace when leading political powers found it convenient to make it so.

6 In fact, after the elections the *Dōshikai* renamed itself the *Kenseikai*, hoping for a fresh start.

7 The associate's name was Nishihara Kamezo, so the loans involved were labelled the Nishihara Loans.

8 Katō's *Kenseikai* supported the Siberian Expedition, hoping to turn that support into an alliance with the army and a way back into the prime ministership. Since Terauchi had become Prime Minister to drive out Katō, the division over the expedition was doubly disturbing.

9 Of particular importance was the Chinese Eastern Railway, the short cut across Chinese territory to the Trans-Siberian Railway's terminus at Vladivostok. Technically the Chinese Eastern was under Russian control, but local Chinese forces had used the outbreak of the Russian Revolution to operate it themselves.

10 Gotō was a career bureaucrat who, as Terauchi's Home Minister, had engineered the crushing defeat of Katō's *Dōshikai* in early 1917.

11 Yamagata was the last surviving original *genrō*. He would die in 1922. Saionji had been recognized as a *genrō* by 1918 and would live until 1940, but he never enjoyed the influence that the original *genrō* had.

12 By this time, all of the Czechs had been evacuated.

5
The breakdown of the Meiji system

The early 1920s were critical for Japan's foreign policy. With the end of the First World War in Europe came the collapse of what remained of the old imperial order in East Asia. Japan had to redefine its fundamental relationships with the three major players in that corner of the world. To complicate that redefinition, every one of these players was in some way new to its role in Asia. First, Japan sought to secure America's understanding of its regional pre-eminence in China at the Washington Conference of 1921–2. At the same time, it had to determine how far to press that pre-eminence against the rising power of the Chinese Nationalists, who would firmly establish themselves as China's dominant faction by the mid-1920s. Last, but hardly least, the leaders in Tokyo had to decide what sort of relationship they wanted with the revolutionary Soviet Union, which had successfully asserted its control over Siberia.

This time was critical for another reason: it marked the passing of the last of the Meiji oligarchs and, consequently, the end of their ability to maintain a reasonable consensus over Japanese foreign policy. In early 1922, Yamagata Aritomo, last of the great *genrō*, died. Saionji Kimmochi, titled a Prince in recognition of his latest service at the Paris Peace Conference, became a sort of acting *genrō*, but he lacked the influence of the original senior statesmen, who had been, after all, nothing less than the founders of modern Japan. Saionji also lacked the skill to bring together the political leaders of Japan, the military, the bureaucrats of the civilian ministries, and the parties in the Diet, to form a basic agreement around what sort of foreign policy Japan ought to have. Hara Kei did have the influence, power, and skill to form such a consensus, but he was murdered in November 1921. No one, from the political parties, the bureaucracy, nor even the military, would be able to unite Japanese politics or policies for the next twenty-four years. So Japan had no unified, or even consistent, foreign policy from the end of the First World War to the beginning of the second. The results would prove disastrous.

The impact of Hara's death was felt immediately within Japan's delegation to the Washington Conference. Admiral Katō Tomasaburō had counted on Hara's influence to avoid difficulties with hardliners inside the army, navy, and political parties who opposed an accommodation with the United States and Britain. The politically inept Takahashi Korekiyo had succeeded Hara as head of the *Seiyūkai* party, and the *Seiyūkai*'s political rivals sniffed easy prey. These rivals did not necessarily oppose good relations with America, but they knew that if they could portray Takahashi and Katō as surrendering Japan's sovereign rights at Washington, the *Seiyūkai* would be doomed in the next elections.

Their first attempt came with the tonnage ratio question. At Washington, Katō had accepted the American proposal allowing Japan 3 tons of battleship for every 5 tons that Britain and the United States could have. Inukai Tsuyoshi and Gotō Shimpei, ambitious political leaders who had been snubbed by the *Seiyūkai*, complained that the ratio was an insult to the principle of Japan's equal sovereignty with any Western nation. They even hinted that the inferior ratio was the result of Western racist sentiment against Japan. Admiral Katō was more concerned with the realities of Japan's defence. He saw that with no treaty at all limiting battleship construction, the Americans could easily build 12 or 15 tons of warships to Japan's 3 tons. Katō also argued that the concession he had won from the West – that the powers would not fortify their Pacific possessions (except for Britain's Singapore, America's Pearl Harbor and of course Japan's home islands) – did far more to protect Japan than insisting on equal tonnage could.

Inukai and Gotō were more interested in weakening the *Seiyūkai* than protecting Japan. They seized upon Admiral Katō's admission that a slightly higher tonnage ratio in Japan's favour (perhaps 7 Japanese tons to 10 American or British ones) would be desirable and demanded that he win this as a concession from the West. They also stretched the definition of 'Japan's home islands' in the non-fortification agreement by arguing that the Bonin island chain well south of Japan ought to be considered 'home' islands.

When these demands reached Admiral Katō in early January 1922, he became outraged. He agreed to have the West allow Japan to complete a large, modern battleship instead of scrapping it,[1] but he refused to try to modify the ratio to allow Japan more ships. When Inukai and Gotō continued to press for the large definition of 'home' islands, Katō coldly threatened to resign as delegate, a step that would have ended Japan's participation in the conference altogether. Inukai and Gotō then threatened to take their case to the public. Katō won this test of wills, since Inukai and Gotō calculated that if they were blamed for sabotaging Japan at Washington their political futures would be harmed. They backed

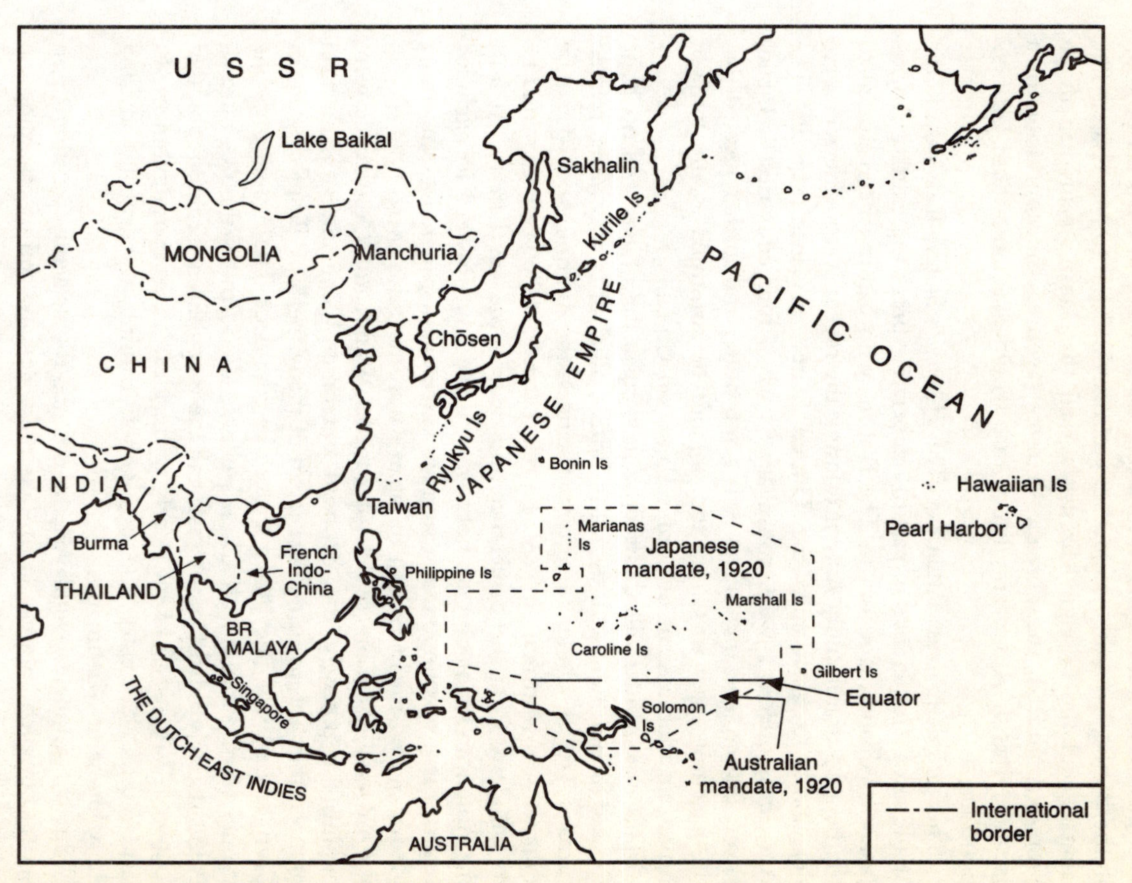

Map 5 Central and Western Pacific 1921

down and Japan agreed to the Five-Power Treaty[2] sharply limiting the battleship navies of the world.

There was no disagreement within Japan over its wider relationship with the West concerning Pacific affairs. Inukai, Gotō, Admiral Katō, and Takahashi wanted the West explicitly to recognize Japan's position in the Pacific, from the conquest of Germany's islands to Manchuria. The military and the Foreign Ministry agreed that Japan's gains had to be protected. Indeed, the Foreign Ministry favoured the conclusion of a Japanese–British–American alliance that would guarantee those Japanese positions.

The problem was America. The United States had called the Washington Conference in part to deliberately break the existing Anglo-Japanese Alliance, not join it. The British, who had just been treated to the benefits of a friendly America during the recent war in Europe, were not about to antagonize the United States on Japan's behalf. The Americans offered a weak consultative pact of four countries – Britain, Japan, France, and themselves – as a substitute, and further insisted that the pact not apply to China and, so, not be construed as any form of American endorsement of Japan's claims on China.

This was a bitter pill to swallow, but Japan did. There was no hope for any foreign support if even the British, allies of twenty years, refused to back Tokyo. Under these circumstances, the existing Anglo-Japanese Alliance was worthless anyway. To reject the Four-Power Pact would be to withdraw from the conference and destroy the naval treaty. At least the American stipulation cut both ways: if the Four-Power Pact excluded China, there was hope that a separate agreement, specifically concerning China, indeed might recognize Japan's wartime gains there.

Of course, the Chinese delegates to the Washington Conference were hoping for precisely the opposite outcome. Confident of American support, they meant to deprive Japan of many, perhaps all, of its gains in China since 1914. They had reason to be confident. The American public strongly sympathized with China's plight. Woodrow Wilson's Fourteen Points strongly aided China's case, and the Republican administration that followed Wilson's had kept closely to his ideals in its own diplomacy. In fact, Wilson's Secretary of State, Robert Lansing, was actually advising the Chinese at Washington.

But the Chinese overplayed their hand. By stridently insisting upon Japan's complete submission to their every demand and by regarding the surrender of all foreign rights (such as extraterritoriality) as a matter of justice, not negotiation, they alienated the powers at Washington, including the United States. The resulting Nine-Power Treaty concerning China was much more favourable to Japan's valued rights in China than Tokyo had first expected.

That Nine-Power Treaty bound all powers not to enlarge their positions and rights in China in the future, but it did not challenge positions and rights gained in the past, even since 1914.[3] The treaty did not explicitly guarantee those positions and rights dearest to Japan – in Manchuria – but the United States and Britain assured Japan that they would interpret the treaty as not contrary to those positions and rights. Did that mean, Japan asked, that the 1915 renewal of Japan's lease in south Manchuria (for ninety-nine years!) was safe under the treaty? It was.

With the major issues of battleship limits and Pacific and Chinese questions resolved, Japan wrapped up two other nettlesome problems at Washington: Shantung and Siberia. Japan was willing formally to surrender most of its rights in Shantung. Its troops would leave Shantung and all of the old German leased territory would revert to Chinese sovereignty. Japan would sell the (formerly German) Tsingtao–Tsinan railroad to China, though a Japanese national would manage the railroad for five years. Japan was willing to give these up in exchange for China's agreement not to hinder Japan's growing economic presence, from coal mines to textile mills, in Shantung generally. In essence, Japan was proposing to take the first step toward changing its position in China from one based on imperial rights to one founded on investment, trade, and mutual prosperity.

This policy delighted the leaders of the *zaibatsu*, who would enjoy a decade of pre-eminence in the 1920s. Their investment in China, primarily in textiles, was already significant. It would grow materially during the decade, though in comparison to direct state investment in Manchuria it remained rather small.

Hardliners, such as the army, Inukai, and Gotō, viewed this transition from imperial rights to mutual prosperity with misgivings. They feared that the Chinese merely meant to use the rhetoric of trade to drive Japan off the continent entirely, including the prize of Manchuria, a region of vital strategic importance in any rematch with the Russians, and one increasingly productive economically.[4]

These fears were exaggerated, but only somewhat. In the early 1920s, few Chinese expected to drive Japan out of Manchuria. But they did hope that the Americans would force Japan out of Shantung entirely. Again they miscalculated. Japan argued that the Paris Peace Conference guaranteed some Japanese presence in Shantung. To overturn one Paris settlement threatened to bring all others up for reconsideration. No power, not even the United States, was willing to open so large a Pandora's box, so the pressure instead was on China to reach an accommodation, which was achieved on Japan's terms.

Likewise, accommodation was in order over Siberia. Hara had ordered the Imperial Army to evacuate Siberia in May 1921, but the army had been dragging its feet since it wanted something concrete to show for its three-year exercise. This became harder after Shidehara Kijūrō – a fast-rising diplomat who strongly believed in cooperation with the West and China – publicly declared at the Washington Conference that Japan would fully withdraw from Siberia. The army balked at first, but agreed to the pull-out after the Foreign Ministry secured concessions from the Soviet Union allowing Japan to develop half of the oil fields and coal mines of northern Sakhalin island.

The Washington treaties would bring nearly a decade of stability to Japan's relations with the West, no mean feat in the tumultuous twentieth century. But the West itself was stable in the 1920s, not so China. There, kaleidoscopic chaos guaranteed further difficulties as Japan had to determine which faction, if any, to support in order to safeguard its rights and positions on the Asian mainland. Before Japan could make any decision about China, however, its leaders had to decide among themselves who would govern Japan. This was no easy task in the early 1920s.

It was clear to his friends and foes alike that Takahashi was a failure as Prime Minister. Takahashi proved unable to keep his *Seiyūkai* together, much less the nation, and resigned in June 1922. On the other hand, Hara had been so successful in promoting *Seiyūkai* strength that the other parties, Katō Takaaki's *Kenseikai* and Inukai's *Kakushin kurabu*, were much too weak in the Diet for their heads to lead the nation. Besides, Katō had a bad reputation in the West as a result of his authoring the Twenty-one Demands against China in 1915. He was hardly a sound choice as the premier to implement the Washington treaties.

Instead Admiral Katō became Prime Minister with *Seiyūkai* support. The navy man was an ideal choice for a Japan that favoured close ties with the West, but he died in August 1923. Since the *Seiyūkai* was still split and its rivals still weak, another navy man, Admiral Yamamoto Gonnohyōe,[5] became Prime Minister.

The Yamamoto Cabinet might have pursued a most interesting foreign policy. Yamamoto himself had respected Admiral Katō and supported the Washington treaties. But Yamamoto also wanted to promote domestic unity, so he proposed to form a cabinet in which the bureaucrats, military, and all political parties would share power. This idea was popular with no one. The resulting cabinet was weak and divided. It was also fated to be short, for in September 1923 a titanic earthquake and accompanying fire destroyed much of central Tokyo. More than 100,000 people were killed and over half a million houses destroyed. In the pandemonium that followed, mobs, sometimes with police and even army assistance, attacked leftists and Koreans. These attacks outraged a young Japanese idealist, who sought justice by attempting to assassinate Prince Regent (soon to be Emperor) Hirohito. Yamamoto and his cabinet resigned in apology for having been in office when such an unspeakable act took place. By January 1924, Japan was again searching for a government.

Nothing had yet changed with the parties, so their leaders were again denied the premiership. This prompted a revolt in the Diet. The dominant *Seiyūkai* split apart. Some members, renaming themselves the *Seiyūhontō* (the 'true *Seiyūkai*'), decided to cast their lot with the army man, Kiyoura Keigo, who had been named premier. But everyone else went into opposition, demanding that the age of the politicians had finally arrived and that henceforth all Prime Ministers ought to come from the parties. So fierce was their opposition that Kiyoura dissolved the Diet and held elections in the spring. Benefiting from the *Seiyūkai*'s split, Katō's *Kenseikai* did well at the polls and, in June 1924, Katō of the Twenty-one Demands became Prime Minister.

Ordinarily, one might have expected Katō to pursue an anti-Western foreign policy and return to a hard line against China. Instead, Katō's chief preoccupations turned out to be domestic reforms. He would lead the fight for voting rights for all male adults and would sponsor labour legislation. His budgets actually reduced the size of the Imperial Army. Katō remained strongly suspicious of communism and hence the Soviet Union, but primarily as a domestic threat from within Japan. Soon after his governmernt signed a treaty with Moscow normalizing relations, the Diet enacted a stiff 'Peace Preservation Law' to curb 'dangerous thought' among the Japanese people.

Much of Japan's foreign policy for the next three years would be in the hands of Katō's Foreign Minister. Shidehara brought to his new post a set of firm convictions about Japan's future in the world. The European war showed that the military path to security led to disaster. The future safety of Japan depended upon its ability to cooperate with other nations and gain prosperity through trade with them. Japan was fortunate that its main export market, the United States, was politically stable and economically robust. Maintaining good relations with Washington was imperative. Even when the Americans insulted Japan with their immigration law of 1924, which virtually barred Orientals from entering the United States, Shidehara refused to protest.

Shidehara was not about to rely upon America alone for Japan's well-being. Japanese trade with, and investment in, neighbouring China had boomed during the European war. Shidehara was determined to build on that success. Trade with the Chinese, he understood, depended upon winning their goodwill. To that end, Shidehara's maiden speech to the Diet in July 1924 proclaimed Japan's strict adherence to the principle of non-interference in China's affairs.

Shidehara's policy was immediately put to the test. By 1924, the Imperial Army was supporting Chinese warlord Chang Tso-lin in exchange for Chang protecting Japan's position in Manchuria. In fact, under the army's influence the short-lived Kiyoura Cabinet had scrapped all pretence of favouring the unification of China under anyone. Chang would guard Manchuria; his rivals would be too weak fighting each other in northern and southern China. This policy worked too well. An emboldened Chang decided to venture out of his Manchurian base to lay claim to all northern China, driving rival warlords into an alliance against him that threatened even Manchuria. Chang appealed to Shidehara – and the Japanese 'Kwantung Army'[6] in southern Manchuria – for assistance.

Shidehara adamantly refused to help Chang. While this stance pleased China and the West, it angered the Imperial Army. As the warlord's situation deteriorated further in the autumn of 1924, the army began to look for allies against Shidehara. It found them in the *Seiyūkai* and *Seiyūhontō*, which began to criticize Shidehara for endangering Japan's rights in China. If these parties withdrew their support for Katō's cabinet, it would fall and Shidehara would be replaced. The showdown came at a cabinet meeting on 23 October 1924. Shidehara held his ground. The dissenting parties backed away from their threat to withdraw support. For the moment, Katō and Shidehara were safe.

As it turned out, Chang was safe too. Unknown to Shidehara, Japan's army had extended additional aid to Chang and had used its agents and money to break up the coalition in China against him. Back in Japan, the army continued to cultivate political allies by deriding Shidehara as an idealistic dreamer whose inaction would pave the way for communism in China. The army pointed out that Shidehara's earlier opposition to the Siberian Expedition had permitted the Bolsheviks to consolidate their hold on the Soviet Union. Shidehara's policies would surrender all Asia to the Reds, the army claimed.

Actually, the army's fears had substance. In early 1924, the *Kuomintang*, China's Nationalist Party, had allied itself with the Chinese Communist Party and gained Soviet support. The combination was potent and popular in China. By the spring of 1925, the *Kuomintang* and Chinese communists were organizing boycotts and labour strikes against foreign businesses in China. Japanese-owned cotton mills were frequent targets for good reason: by the 1920s a large number of Japanese companies, including the *zaibatsu*, had become a major force in the textile industry. These businesses asked Tokyo for protection against the Chinese boycotts and strikes, which sometimes turned violent.

Shidehara refused. He argued that, in the long run, Japan and Japanese businesses would do better if the Chinese were friendly. Sending troops would be counter-productive. Instead, Shidehara appealed to the *Kuomintang* to arrest 'outside agitators' and maintain orderly demonstrations. He also recommended that the Japanese mill owners grant concessions to their Chinese workers to end the strikes and boycotts.

Shidehara's approach was unpopular among Japanese businesses who operated in China and, of course, the Imperial Army and its newfound political ally, the *Seiyūkai*. But it might have worked, except for the unfortunate May Thirtieth Incident of 1925. On that day a group of Chinese demonstrators charged a police station in Shanghai's quarter reserved for foreigners. The British commander ordered the police to fire, killing nine unarmed protestors. The resulting nationwide fury renewed Chinese militancy against all foreigners, including the Japanese.

Shidehara understood that he was in a difficult situation. But he persisted in arguing for conciliation. If Japan, or any foreign power, took a hard line against the Chinese demonstrators, they would simply resort to harsher measures themselves. The result would further weaken any authorities in China and invite chaos – playing into the hands of the Chinese communists. It was better to be reasonable and try to shore up moderate authorities in China. The best way to do that was to agree to an international conference to return tariff autonomy to China.

It was a clever move. The Chinese demonstrators had complained that Chinese-owned textile mills in China could not compete with foreign mills because China could not raise tariffs to protect its own industry. Shidehara's proposal appealed to them greatly. As well, at any tariff conference it would be Shidehara's Foreign Ministry, not the Imperial Army, that would be in control.

Japanese exporters to China were predictably cool to the idea of a tariff conference. If China gained the sovereign right to determine its own tariffs, they would certainly rise, perhaps steeply. Japanese exports would be hurt, possibly ruined. Shidehara, therefore, approached the Peking Tariff Conference cautiously. He was prepared to yield the principle of tariff autonomy, in exchange for China agreeing to keep actual increases minimal, at least against Japanese products. If there were higher tariffs against Western goods, Japan was prepared to accept them.

The conference closely followed Shidehara's script. The powers agreed to grant tariff autonomy to China by 1929. China consented to separate talks with Japan for tariffs on Japanese goods. The West was unhappy with that arrangement, but Western attempts to devise a general set of Chinese tariff rates came to nothing. Shidehara was pleased with the immediate result of the Peking Tariff Conference, but he was disturbed by the reason the Chinese Government did not agree to any general set of tariff rates: by mid-1926 there was no central Chinese authority in Peking. Renewed fighting among the north China warlords swept away even the pretence of an overall government there.

None of this posed any immediate difficulty for Shidehara's overall policy of non-interference in China's political affairs. But it did promise to bring fresh turmoil to those affairs, and renew calls from Japanese in China for protection. The complete breakdown of government in northern China provided a perfect opportunity for the south China-based *Kuomintang* to bid at last for total control of the entire country. Chiang Kai-shek, the *Kuomintang*'s military commander, announced his 'Northern Expedition' to reunify China. It began in early July.

Shidehara's response came in his opening speech to the Diet in January 1927. In a shrewd mixture of principle and practicality, Shidehara observed that it was up to the Chinese to determine their own future without foreign interference – a reference to the *Kuomintang*'s reliance on Soviet support. Japan was willing to assist China in the spirit of co-prosperity and mutual respect. But that co-prosperity was grounded upon respect for Japan's economic rights in Manchuria and for the safety of Japanese to conduct peaceable business anywhere in China.

Shidehara's speech actually pleased the *Kuomintang*, whose members had feared that Japan would resist the Northern Expedition, especially if it entered Manchuria. But Chang Tso-lin saw it differently. He ignored Shidehara, convinced that the Imperial Army would never abandon him in Manchuria in exchange for the *Kuomintang* and their communist allies. So he determined to resist the Northern Expedition himself and reassert his control over all north China in the process. Chang's gamble was cagey, but also risky. He was hoping that the Imperial Army would win complete control over Japan's foreign relations, replace Shidehara, and join him in attacking the *Kuomintang*.

At first, Chang's strategy went well. Although *Kuomintang* troops seized Shanghai, China's largest port city, in March 1927, and took the southern capital of Nanking soon after, they created trouble with all foreign countries in so doing. Unruly *Kuomintang* troops looted foreign businesses and killed foreigners in Nanking, including a Japanese naval officer. A crisis with Japan and the West seemed imminent.

That crisis never arrived. Chiang at once apologized for the Nanking incident and asked for time to impose better discipline upon the *Kuomintang*'s troops. Chiang's pledge was double: he not only wished to prevent future difficulties with the powers by avoiding repetition of the Nanking incident, Chiang also moved decisively to drive the communists out of the *Kuomintang*. His troops confiscated weapons from striking workers beginning in April even as Chiang's top lieutenants went to Tokyo to solicit aid in the new *Kuomintang* drive against the Chinese communists and their Soviet supporters.

Chiang's diplomacy, especially his treatment of Japan, was brilliant. He had addressed every major Japanese concern about his *Kuomintang*, with the important exception of the future of Manchuria. Yet, even there, Chiang was signalling that an accommodation was possible if Japan assisted, or at least did not resist, his consolidation of power over the rest of China. In the spring of 1927, there was a real chance for a long-term, stable peace between Japan and China of exactly the sort Shidehara had been working for.

But by April, Shidehara was no longer Japan's Foreign Minister. He was, instead, the victim of particularly vicious political manoeuvres within Japan, manoeuvres that used Shidehara's own foreign policy as their battleground.

For nearly three years the Imperial Army had worked for Shidehara's ouster. For three years, however, Shidehara and Prime Minister Katō had remained in control and had seen through an ambitious programme of domestic reforms. It was a rare case of policies, even principles, over politics. In the Diet, Katō had no difficulty keeping his own *Kenseikai* in line. But the second largest party, the (still leaderless) *Seiyūkai*, chafed at its junior treatment as time wore on. Members of the *Seiyūkai* knew that the surest way to return to power was to recruit a bold, charismatic leader from outside the party's regulars. The army's Tanaka Gi'ichi seemed the perfect choice. He became leader of the *Seiyūkai* in April 1925.

Tanaka was a big, blustery man with strong views. He hated leftist radicalism, especially communism. Tanaka and his *Seiyūkai* had been willing to agree to Katō's proposal allowing all adult males the right to vote, but only if the Diet also passed an onerous Peace Preservation Law granting the Japanese Government sweeping powers to suppress dissent. Tanaka had strongly favoured the Siberian Expedition, and he was constantly critical of Shidehara's conciliatory treatment of China, arguing that it would just encourage the growth of communism there.

Under Tanaka's leadership, the *Seiyūkai*'s attacks on Shidehara's 'weak-kneed' diplomacy grew increasingly strident. In January 1926, the ailing Katō died, giving Tanaka and his *Seiyūkai* the opportunity to break away from the cabinet. A year's manoeuvring to unseat the *Kenseikai* Cabinet (under the amiable but weak Wakatsuki Reijirō) finally paid off in April 1927, when Tanaka became both Prime Minister and Foreign Minister in the new, *Seiyūkai* Cabinet.

Tanaka came to power on the basis of promises to reverse Shidehara's policies. His slogans loudly proclaimed that Japanese nationals should be protected by Japanese troops, not Chiang's worthless words. As importantly, he brought into his cabinet leaders, such as Vice Foreign Minister Mori Kaku, who were even more anti-Chinese than he.

Tanaka swiftly moved to reverse Shidehara's conciliation of the *Kuomintang*. In late May, he announced that 2000 Japanese troops would be sent to Shantung, just as *Kuomintang* soldiers were advancing there. In

June and July he held 'The Eastern Conference,' a media event that Tanaka used to trumpet his determination to defend Manchuria to the utmost. Behind the scenes, Tanaka spoke of merging the administrative apparatus of Korea and Manchuria. This was a highly provocative step that would make Manchuria a formal part of the Japanese Empire and remove it forever from Chinese control, but Tanaka was prepared to take it. In July he named two friends, Yamamoto Jōtarō and Matsuoka Yōsuke, President and Vice-President of the South Manchurian Railway. They – not Japan's Foreign Ministry, which still had many Shidehara supporters – would negotiate with Chang Tso-lin for additional Japanese rail lines throughout Manchuria. By the end of 1927, Japan's foreign relations were in the hands of the hardliners: the army and the *Seiyūkai*, united under Tanaka.

Tanaka was able to undertake these bold measures to secure Manchuria because, for a time, Chiang Kai-shek was too preoccupied in ousting the communists from the *Kuomintang* to worry about Japan. Yet Chiang's own consolidation of power posed a problem for Tanaka. Chiang appeared to be the best weapon China (and Japan) had against communism in China. It made no sense to weaken Chiang's *Kuomintang* – so long as Chiang agreed to stay away from Manchuria and possibly Shantung, where Japan retained some interests as a result of its occupation since 1914. Tanaka and Chiang spoke with each other in Tokyo in November. No formal agreement resulted, but Chiang hinted that he might stop his forces outside of Manchuria and would not force battle with his opponents in Shantung, eliminating the need for Japanese forces to guard Japanese residents there. Tanaka suggested that he would not contest Chiang's rule throughout the rest of China. One month later, Chiang's *Kuomintang* resumed its Northern Expedition under these understandings.

By the end of 1927, then, Tanaka's diplomacy toward China did not look remarkably different from Shidehara's. To be sure, Tanaka was much more insistent on much more Japanese control over much more of Manchuria (and Shantung) than Shidehara had been. But an accommodation with Chiang appeared possible, even likely.

Once again, however, Japan's politics interfered with its foreign policy. The rivals of Tanaka's triumphant *Seiyūkai* had combined their forces to form a united opposition party, the *Minseitō*, and forced an election for mid-February 1928. It was a historic occasion: the first election held in Japan in which all adult males could vote. For that reason, both parties outdid themselves in seeking issues calculated to appeal to the common man. The *Seiyūkai* effectively labelled itself the party of protection, against radicalism and foreigners, at home and abroad. A prominent plank of its electoral platform called for safeguarding Japanese in Shantung.

Tanaka and his *Seiyūkai* barely clung to power with a one-seat edge over the *Minseitō* after the election results were in. So Tanaka did not have much freedom to deviate from his party's electoral promises, especially prominent ones. In April, *Kuomintang* forces were entering Shantung. Chiang declared his guarantee of safety for Japanese nationals. But Japanese businessmen and residents there were worried, so Tanaka sent 5500 troops to Shantung anyway. Britain, America, the Soviet Union and, most of all, Chiang, strongly protested.

Tanaka had not betrayed Chiang. At least, he had not intended to. But the Imperial Army was determined to teach Chiang a lesson. Japanese troops marched straight into Shantung's capital city, Tsinan, to confront *Kuomintang* forces as they arrived. Almost inevitably, shooting began.

Chiang moved at once to resolve this 'Tsinan Incident'. He offered to pull his troops out of the city, but the Imperial Army wanted more: Chiang was to apologize and accept guilt for causing the incident in the first place. This demand provoked heated debate within Tanaka's cabinet. The Foreign Ministry argued that Chiang had behaved well, even graciously. The army insisted upon giving the *Kuomintang*'s leader an ultimatum to speed along his apology and acceptance. *Seiyūkai* leaders reminded Tanaka of their recent pledge to the voters to protect the Japanese in Shantung. Tanaka approved the army's ultimatum, which went to Chiang on 7 May. One day later, 15,000 more Japanese troops departed for Shantung. Once again, the Imperial Army had demonstrated that it, and no one else, controlled Japan's policy toward China.

Chiang still was more interested in consolidating his power within China rather than fighting Japan. He accepted the Imperial Army's ultimatum, only to find that the army rejected his acceptance – still more terms would be demanded. While the *Kuomintang* waited to learn them, Japanese forces attacked Tsinan and drove the *Kuomintang* from the city by force.

Tanaka hoped to keep his November deal with Chiang. He wanted to pressure Chang Tso-lin, who was still holding out in Peking and still had pretensions to becoming ruler of north China, to retreat into Manchuria and not challenge the *Kuomintang* in China proper. Chiang's forces would not enter Manchuria, but Tanaka was prepared to allow the *Kuomintang* flag to fly over that region as a face-saving device.

Tanaka's willingness to deal with Chiang even in Manchuria infuriated officers of Japan's Kwantung Army. They thought it was intolerable that a Chinese flag should fly there, and feared that Tanaka's weak policy would lead Chang to make his own deal with Chiang. When Tanaka refused to let them disarm Chang's troops as they retired into Manchuria, a young Japanese officer[7] blew up Chang's personal train as it entered the region, killing him instantly. The officers of the Kwantung

Army were sure that rioting and disturbances would follow Chang's death, creating the perfect excuse for their Japanese forces to seize all Manchuria.

Instead, Chang's assassination backfired. His son, Chang Hsüeh-liang, maintained an orderly withdrawal into Manchuria. But within weeks he learned who had killed his father, and he quickly allied with the *Kuomintang*.

This was doubly bad news, since Chiang had been outraged by Japan's actions in Shantung and Manchuria. He was not ready for an open break with Tokyo, but he did press for the immediate granting of tariff autonomy from the powers (a grant that had been delayed three years earlier by the breakup of the Peking Tariff Conference) and, worse for Tanaka, received it from every nation except Japan. By early 1929, a furious Chiang informed Tokyo that China would enact its own tariffs regardless of existing treaty obligations. If Chiang was going to defy agreements on tariffs, Tanaka feared he was likely to challenge Japan's position in Manchuria, too.

Tanaka's difficulties with Chang and Chiang provided the opposition *Minseitō* with potent political ammunition against him. Even more explosive was the fate of the army officer who had murdered the elder Chang. Saionji, a founder of the *Seiyūkai* and a senior statesman of great stature, pressured Tanaka to have the officer tried for his crime. The Imperial Army insisted that he had acted from understandable impulses and that, even if he had been wrong, he fell under the army's jurisdiction, no one else's. Tanaka, an army careerist and head of the *Seiyūkai*, was skewered between both his bases of power. He would have to alienate one or the other. Understandably, he tried to delay any resolution of the affair as long as possible, but in June 1929 he formally reported to the Emperor that the Imperial Army had not been involved in Chang Tso-lin's assassination. It was a bare-faced lie, but it removed any possibility that the army officer would stand trial. Of course, precisely because it was a lie, as everyone knew, Tanaka was compelled to resign in early July. Once again, the Imperial Army had demonstrated its dominance of Japan's policy in China.

The new *Minseitō* Prime Minister, Hamaguchi Osachi, wasted no time in bringing Shidehara back as Foreign Minister. Shidehara wasted no time in repairing the damage that Tanaka had done to Japan's relations with China and the West. Still certain that Chiang was the real power in China, Shidehara moved to conciliate the *Kuomintang* leader. By March 1930, Japan had agreed to grant China tariff autonomy in exchange for relatively light tariff increases on Japanese goods. In April, when a coalition of warlords challenged the *Kuomintang* in northern China, Shidehara made no effort to assist them. At the same time, after the Chinese communists rose in central China and seized the city of Changsha for ten days, killing many foreigners, Shidehara refused to consider the dispatch of Japanese forces. Nor did he insist upon holding Chiang accountable for allowing such disorders. After Chiang successfully enlisted the young Chang in his cause to put down the northern rebellion, ensuring that Chang's Manchuria would be a part of the *Kuomintang* regime, Shidehara refused to act to protect Japanese interests there.

Shidehara also moved Japan closer to the West. In January 1930, Japan returned to the gold standard. The gold standard had been the mechanism for ensuring stable currency values among the advanced nations before the great European war of 1914–18. The economic chaos of that war prevented many countries from rejoining the gold standard, since such a step meant guaranteeing that their currencies could be freely exchanged for gold. In this way, nations on the gold standard could not arbitrarily alter their currencies' value to achieve advantages in either importing or exporting their goods. In political terms, rejoining the gold standard meant reclaiming great power status (that is, membership in the club of advanced nations) financially. It also represented confidence that the nation's goods were competitive internationally.

Just as important, Shidehara returned Japan to serious discussions with Britain and America regarding limits on naval armaments. The Washington treaty of 1922 had resolved the issue of battleship competition. But the three naval powers had begun rival construction programmes of another type of warship, the cruiser. To settle the cruiser question, their delegations gathered in London in January.

The London Conference quickly deadlocked over the tonnage ratio question. The United States insisted that the same 5:5:3 ratio that governed battleship tonnage allowed to each power be extended to cruiser tonnage. The Japanese remembered that they had agreed to the 5:5:3 ratio at Washington only after receiving the valuable concession of the West not fortifying its Pacific island possessions. The Americans were offering nothing in exchange this time. Moreover, many in Japan, especially in the Imperial Navy, had come to view the 5:5:3 ratio as a perpetual symbol of Japan's inferiority to the West. The boldest, such as Admirals Suetsugu Nobumasa and Katō Kanji (not related to the Washington Conference Admiral Katō Tomosaburō), preferred outright parity. Even the moderates demanded a better ratio of 10:10:7.

The impasse was broken by a compromise worked out with great effort and difficulty by a Japanese and an American delegate. Technically, the 5:5:3 ratio was still in effect. Practically, the United States would refrain from building all of the cruisers allowed by the ratio until 1938. As a result, for most of the decade Japan would have a *de facto* ratio of 10:7 compared to the Americans. Only by 1938, with the completion of the final American cruiser allowed, would that ratio drop to 10:6. Before then, there would be another naval conference that, hopefully, would settle the difference permanently.

Japan's delegates to the London Conference met in mid-March to consider whether to recommend the compromise to their government. The chief delegate, former Prime Minister Wakatsuki, agreed that it should be. In Tokyo, Shidehara approved, but Prime Minister Hamaguchi wanted to know the Imperial Navy's opinion before proceeding.

The result was civil war inside the navy. The Navy Ministry was in the hands of disciples of the departed Admiral Katō Tomosaburō, statesman of the Washington Conference. Predictably, these men valued an agreement with the United States that avoided the prospect of an arms race and worsened relations. But the Naval General Staff, in charge of actual fleet command and operations (as opposed to budgetary, diplomatic, and political concerns), was in the hands of Katō Kanji and Suetsugu. They were 10:7 diehards.

Ordinarily, dissent from the General Staff was troubling but hardly critical. But 1930 was different. The Imperial Navy had learned from the army. It was organizing its own political alliances to allow it to pursue its own policies. Specifically, Katō Kanji had lined up the support of Fushimi Hiroyasu and Tōgō Heihachirō. Fushimi was a top military leader and, perhaps more to the point, an imperial prince – a blood relative of the Emperor and thus with easy access to him. Tōgō had commanded the Japanese fleet at the crushing victory of 1905 over the Russians in the Tsushima Straits and was a national hero who commanded immense respect. Both recommended that Japan insist upon a 10:7 ratio or withdraw from the conference. To make matters worse for the proponents of the London treaty, the *Seiyūkai* [8] eagerly associated itself with the navy's demands.

Hamaguchi himself had a crucial decision to make: should he proceed with adopting a London naval treaty on the basis of the Japanese–American compromise, or avoid a fight within Japan by having the London Conference break up without agreement?

It was clear that any fight would extend to the Diet. Even before the split within the navy, *Seiyūkai* leader Inukai Tsuyoshi had aligned his party with the diehards. On the other hand, Japan's constitution did not vest the power to ratify treaties in the Diet anyway. That power remained with the Emperor, who by custom was to follow the recommendation of the secretive Privy Council, where (hopefully) Saionji's voice for moderation would carry weight.

Hamaguchi decided to press for a treaty based on the compromise. In part, he was buoyed by his *Minseitō*'s strength in the Diet. After elections in February, it enjoyed an absolute majority. The *Seiyūkai* could rant all it wanted. It lacked the influence to throw Hamaguchi out of office. In part, Hamaguchi was moved by budgetary concerns. He wanted to restore fiscal prudence to Japan. Indeed, he had to as part of his policy to return Japan to the gold standard. If the London Conference broke up, an expensive naval construction programme was certain to follow.

More importantly, Hamaguchi was absolutely convinced that the treaty was imperative for Japan's long-term security. To withdraw from the conference would shatter hopes of Japan moving closer to the West – the keystone of Shidehara's diplomacy. Although it would be difficult to give up many of Japan's imperial rights in China, Shidehara and leaders like him believed that this was necessary for Japan in the long run. Moreover, the Emperor agreed. He had privately informed Hamaguchi that a treaty was highly desirable. Hamaguchi would not press for ratification at once, to allow diehard tempers to cool, but he would use the time to ensure a favourable vote within the Privy Council. On 1 April, the Japanese Government approved the compromise.

Three weeks later a new Diet session opened in Tokyo. There, Admiral Katō Kanji precipitated a full-blown constitutional crisis by claiming that the treaty violated the military's 'Right of Supreme Command' (*tōsuiken*). Smelling blood, an eager *Seiyūkai* joined in the attack on Hamaguchi, Shidehara, and the treaty.

The Japanese Constitution of 1889, or Meiji Constitution, indeed granted the military, not any civilian, the right to command the armed forces of Japan. However, Katō's interpretation of *tōsuiken* was far broader. He claimed that the navy and army had a constitutional right to determine the size of their forces, not just the power to command and operate those forces given them by the government's budgets. Because the London treaty imposed limits on the size of the navy, it was unconstitutional. Katō's case was weak legally, but the *Seiyūkai* leaders knew that it was terrific politics. Likewise, the Imperial Army, which normally might have steered clear of such controversy, was willing to take strong measures to oust Shidehara and his conciliatory policy toward Chiang, so it joined the navy in the *tōsuiken* controversy.

Shidehara assumed his usual confrontational stance. In his opening address to the Diet, he defended the London treaty as both responsible and reasonable and condemned his opponents for lacking confidence in themselves and in Japan. An outraged Katō termed the speech, 'vomit,'[9] and the fight was on.

It could not be settled in the Diet, which adjourned acrimoniously in mid-May. Katō and his allies took their case to the public with speeches and circulars that denounced Shidehara as a near-traitor. But Hamaguchi had his own ace to play. He had his Navy Minister delay returning to

Japan until after the Diet session had ended. He arranged for a spectacular reception for the Navy Minister's arrival, knowing that the minister would do his duty as a cabinet member by supporting the treaty, and refuting Katō's charge that the treaty betrayed Japan and infringed upon *tōsuiken*.

It was a good plan, but it went awry immediately. The Navy Minister arrived to a thronging crowd, but one man tried to stab him. As he travelled to Tokyo, a junior officer on the Naval General Staff committed suicide on a train similarly bound. The minister arrived in the capital to be greeted by leaflets branding him a traitor. He then met a cold Katō, who insisted that he deliver a message to the Emperor opposing the treaty. When the Navy Minister refused, Katō threatened to present it himself and then resign. Katō carried out both threats on 10 June.

Katō's extreme action forced the Navy Minister to go to the Emperor himself to repair the damage. He was relieved to find that there was none. Emperor Hirohito calmly returned Katō's message, unread, noting that Katō had acted improperly in delivering it. The Navy Minister then returned to his office, accepted Katō's resignation from the Naval General Staff, and ensured that Katō's replacement favoured the treaty. He also circulated a memorandum reminding all naval officers that the determination of the size of the fleet was the Navy Minister's, not the Chief of Staff's, responsibility.

The treaty still had hard going, though. The army and navy general staffs formally reported to the Emperor that the treaty was acceptable only as an interim measure and only if the navy were built to the absolute maximum strength allowed in the treaty, including new investments in naval air power. Prime Minister Hamaguchi was forced to agree to a large increase in the navy's budget. The treaty next went before the Privy Council, which had several vehement opponents of Shidehara. They were determined to use the ratification hearings as a way to attack the Foreign Minister's diplomacy. As those hearings became increasingly acrimonious, gleeful *Seiyūkai* leaders predicted the fall of Hamaguchi and Shidehara within days.

They were wrong. Hamaguchi, in a display of determination rare for a prime minister, made it clear that he would recommend the treaty's approval to the Emperor even if the Privy Council voted it down. Faced with the prospect of a constitutional crisis, the privy councillors grudgingly elected to back the treaty. On 2 October 1930, the Emperor ratified the London Naval Treaty.

Hamaguchi was able to relish his victory for 6 weeks. On 14 November, he was gunned down by a rightist who supported the military. He would hold on to life until the next summer, but had to resign in April 1931 due to his grave injuries.

Hamaguchi's loyal Navy Minister was long gone by then. He had resigned immediately after the treaty's ratification, knowing that he could not continue to serve in such a bitterly divided service. Not content with his departure, the followers of Admiral Katō pursued a vendetta against all naval officers who had supported the London treaty until, by 1933, none was left within the navy. In that year, Katō's right-hand man, Suetsugu, became commander of the combined fleet and, as ominously, the Naval General Staff explicitly won from the Navy Ministry the right to determine the size of the fleet in the future. The *tōsuiken* issue was not going to go away. The outlook for any naval conference in 1936 was not bright, nor were Japan's chances for continuing good relations with the West. In many respects, Hamaguchi, Shidehara, and those other Japanese who favoured good relations with the West had won the battle but lost the war.

The London treaty fight also deeply affected the Imperial Army. The army had never been fond of Hamaguchi's determination to restrict military spending. To do so through the device of a treaty – under the authority of the Emperor – was doubly alarming because it was long term and irreversible. Then too, the army had vehemently disagreed with Shidehara over Japan's Chinese policy. Many army officers, therefore, privately rejoiced when Hamaguchi was shot.

Indeed, many were ready to proceed much further. They favoured a full-blooded military *coup d'état* to seize power from the 'weak' *Minseitō* politicians. Their attitude was hardly a secret. For a time, while Hamaguchi was dying, Saionji actively considered naming Army Minister Ugaki Kazunari as Prime Minister precisely to head off such an attempt.

Saionji had reason to be concerned. By 1930, many Japanese in the military and many in civilian life felt that Japan had become weak and soft. The Great Depression, which had started in the United States a year earlier, was reaching Japan. Thousands were losing their jobs in the cities. Even the farmers were hurt as prices for their crops fell steeply. The price of rice, for example, principle crop of a still overwhelmingly rural Japan, was cut in half. The price of silk, Japan's leading export and a key source of supplemental income to many farmers, fell even more. Many blamed the Western-style capitalist economic system that Japan had adopted. They imagined the peaceful, untroubled lives of their grandparents (which in actuality had been not peaceful at all) and longed for a return to those days. In this environment, it was easy to sympathize with the demands of young fanatics in the army that Japan be swept clean by removing the big businessmen and the spineless politicians who served them, and ushering in a 'Shōwa Restoration' that would reorder Japan the way the Meiji Restoration had sixty years earlier.[10]

Within Japan, the army wanted fundamental political changes. But its most immediate worry was preserving its hold over Manchuria. By the spring of 1931, even more senior, more moderate officers such as Ugaki felt that the situation there had become so urgent that direct action was needed. Once the army secured Manchuria, it could turn its attention to fundamental reforms within Japan. Indeed, since Manchuria would be an army province, its governance could serve as a useful experiment in the army's future plans for Japan, an experiment in direct military control of the economy.

Ordinarily, the army would have had little chance to attempt such bold steps. But 1931 was not ordinary. As the Great Depression deepened, more cries rose for radical change. Many *Seiyūkai* leaders were willing to embrace the army's ideas. The *Minseitō*, which might have been a force for stability, had been weakened by the London treaty fight and Hamaguchi's death.

As it was, the less forceful Wakatsuki succeeded Hamaguchi as Prime Minister in April. He had an inkling of the army's plan to take action in Manchuria, and even suspected (rightly) that the army had been behind a rebellion against Chiang's *Kuomintang* in northern China in early 1931. Wakatsuki tried to restrain the army, but he was helpless. When he confronted his own Army Minister (Minami Jirō), he received bland assurances that the army had been involved in nothing untoward. A worried Wakatsuki then had his Home Minister (Adachi Kenzō) issue an order barring military officers from speaking on political or diplomatic affairs.

This was weak medicine that did nothing to stop or even slow the army's rush to seize Manchuria. Indeed, Adachi's posturing helped convince the army's officers in Manchuria to accelerate their plans and created sympathy for them throughout the army, which resented Wakatsuki's attempts at restraint. Conveniently, a Japanese army captain named Nakamura Shintarō had disappeared while travelling in China. Searching for him could provide a perfect excuse for armed action.

The Nakamura incident also gave the army useful political ammunition against the *Minseitō*. Army speakers, in open defiance of the Home Ministry's order, found large audiences throughout Japan eager to listen to attacks on Shidehara's weakness toward the Chinese. By 1 September, the *Seiyūkai* had joined the assault with party chief Inukai, in speech after speech, personally calling for a hard line with the Chinese over Nakamura's disappearance and a speedy resolution of the Manchurian problem.

The *Minseitō* was paralysed in the face of these attacks, especially since Shidehara stubbornly refused to give in to them. He insisted upon continuing calm, deliberate negotiations with China over the Nakamura incident. He continued to believe that a patient approach with the Chinese would serve Japan well in the long run. But Shidehara, like Wakatsuki, had no way of bringing discipline to an army that believed otherwise.

Instead, this task fell to the Emperor and his advisers. The aged Saionji, in a rare display of leadership, arranged to have the Emperor ask the army and navy ministers to maintain order in their ranks. Intervention by the Emperor, personally, in any affair of state was reserved traditionally for matters of the gravest importance. Shocked and mortified, the army's leadership moved to send a senior officer to Manchuria to put all plans to seize that region on hold. Meanwhile, Shidehara moved swiftly to settle the Nakamura incident with Chiang Kai-shek and his Manchurian ally, Chang Hsüeh-liang.

These last-minute actions simply convinced the Japanese officers in Manchuria to act before they were ordered home. They resented the Emperor's interference, believing that he had acted only upon the unwise advice of the cowardly Shidehara. Late on the night of 18 September, they blew up a section of the South Manchurian Railway at Mukden. Blaming the Chinese forces in the vicinity for the explosion, the Japanese attacked immediately. Shidehara's diplomats in Mukden attempted to begin negotiations only to be blocked by an army officer with his sword drawn.

Shidehara had no better luck in Tokyo. At a hastily called cabinet meeting, Shidehara bitterly attacked the army for provoking hostilities. However, as the army had known all along, once Japanese troops were involved in fighting, Shidehara would find it difficult to advocate abandoning them. In fact, the Japanese army in Manchuria had planned a very broad campaign to deliberately stretch its forces to the limit, making reinforcements necessary to save Japanese lives. Already the commander of Japanese forces in Korea, General Hayashi Senjūrō, had issued the orders having Japanese forces cross into Manchuria – without even waiting for the cabinet's approval.

Preparing for a showdown with Shidehara and Wakatsuki, the army determined that its leaders would resign their posts if necessary to bring the cabinet down. This drastic step proved superfluous, as Wakatsuki, worn down by the months of ceaseless assaults on his *Minseitō*, caved in and agreed to approve the army's reinforcement. Wakatsuki was able to win the army's consent not to extend hostilities to northern Manchuria, but only because the army's leaders in Tokyo feared possible Soviet intervention if Japanese troops approached the Siberian border.

Those leaders themselves were shocked when the Kwantung Army – Japan's army in Manchuria – moved north anyway, claiming its own 'right of supreme command'. But they had little recourse except to approve northern operations in the end, especially after Wakatsuki showed no desire for further skirmishing with the army, in Manchuria or Japan.

Part of Wakatsuki's weakness stemmed from the latest attack on his cabinet. In late October a group of army colonels had proposed using the Manchurian crisis to launch a *coup d'état* within Japan. Their plot went nowhere, vetoed by the army's leaders, but rumours of it circulated throughout Tokyo, further intimidating a weary Wakatsuki and increasingly beaten Shidehara.

In part, Wakatsuki and Shidehara realized that the fundamentals of their foreign policy were already in ruins. The West, through its League of Nations, had insisted that the fighting in Manchuria be resolved by negotiations that would end in the reimposition of order – and Chinese sovereignty – over Manchuria. No political power in Japan, except for the *Minseitō*, would accept any solution short of a fundamental reordering of Manchuria to guarantee Japan's rights. In short, Manchuria was not to be part of China at all. The cabinet's approval of northern operations, far beyond the South Manchurian Railway and existing Japanese rights there, was its admission that the Manchurian issue would be resolved to the army's satisfaction, even at the cost of a deep breach in its relations with the West and China. Shidehara's diplomacy was over. Japan's Kwantung Army continued to attack in Manchuria and ignored the League of Nations' attempts to begin negotiations. By early December, the League – the West – had given up, calling for a withdrawal of Japanese forces from Manchuria, but setting neither time limits nor punishments if Tokyo refused. Instead, the League agreed to send an investigative mission that hopefully would provide the basis for negotiations over the future of Manchuria once tempers had cooled.

By the time the League of Nations began its investigation in early December, Wakatsuki and Shidehara were gone. With them disappeared any remaining chances for a negotiated settlement. The principle cause for their fall was their inability to control the Imperial Army and its foreign policy. But they also were the victims of the Great Depression, in two ways. First, the terrible distress within Japan thoroughly discredited the *Minseitō*. As well, the keystone of the *Minseitō*'s financial and foreign policies – putting Japan back on the gold standard in 1930 – backfired spectacularly in December 1931, when Great Britain left the gold standard. As a result, British goods became less expensive than Japanese ones in global markets, and Japanese exports, already hard hit, took a further pummelling. Now even the business community within Japan, which had favoured the *Minseitō*'s desire for closer cooperation with the West and access to the West's markets, turned away. The rush by both Britain and America to raise protective tariffs around themselves only further embittered Japanese business leaders and convinced them that the *Minseitō*'s policies had been wrong all along.

But who should succeed Wakatsuki and Shidehara? This was not an easy question. The *Minseitō* still held a solid majority in the Diet. Just as importantly, the leaders of both major parties were badly split over whether to collaborate with the most popular new force in politics of the day: the army. Saionji, who had inherited the prerogative of the *genrō* in selecting the next prime minister, was not about to let the army take power. So he chose *Seiyūkai* chief Inukai as Japan's next premier.

Inukai's selection blocked the army from the prime ministership, but it hardly limited the army's growing political influence. Several *Minseitō* leaders immediately bolted from their party to set up a new political association that flirted with the army. Within Inukai's own *Seiyūkai*, powerful chieftains openly cultivated senior officers. Keenly aware of his vulnerability, Inukai himself curried favour by appointing General Araki Sadao, darling of the Kwantung Army and would-be coup-makers, as his Army Minister. Mori Kaku, firebrand critic of Shidehara's diplomacy of accommodation, became Chief Cabinet Secretary, a powerful political post. The new Foreign Minister, Yoshizawa Kenkichi, was known as a quiet man who would not make waves; that is, he would not oppose the army's foreign policies.

Inukai also moved to entrench his *Seiyūkai* in power in the Diet. The elections of February 1932 revealed the masses' disenchantment with the old *Minseitō* as the two parties reversed positions. Now the *Seiyūkai* had a commanding majority in the legislature. It was, in essence, a popular endorsement of the army's position.

These developments promised a clear path for the army in Manchuria. By the end of December, the Kwantung Army opened an offensive against Chang's headquarters in southwestern Manchuria. Chang elected to retreat without a fight. He had been heavily criticized by southern Chinese leaders (themselves conveniently far away from the Kwantung Army) for failing to resist Japan from the beginning, and many northern Chinese warlords were eager to see Chang eliminated, or at least humiliated. By February, the Imperial Army had taken all of Manchuria. There only remained the rear-guard action of waiting for the League of Nations' investigative commission to arrive and block it from starting any negotiations between China and Japan that would weaken Tokyo's new grip on Manchuria. The question of Japan's rights and positions in Manchuria had finally been solved – by force.

But the question of those positions in the rest of China was as pointed as ever. Many ordinary Chinese were furious at their leaders' inability to stand up to Japan. Shanghai, with many Japanese textile mills and trading houses, became the focal point of Chinese protest. Speeches, demonstrations, boycotts of Japanese goods – these were not new. But the organization of an anti-Japanese militia was. In response, Japanese residents of Shanghai began organizing their own para-military groups and called on Tokyo to provide protection. Inukai, following in Tanaka's footsteps, rushed nearly 2000 marines to Shanghai by late January.

After a Japanese Buddhist priest was killed by a Chinese mob, the marines landed, determined to show that the Imperial Navy could field fighters every bit as formidable as the army's in Manchuria. Spoiling for action, the marines got it at midnight on 28 January as they ran into tough Chinese veterans of the *Kuomintang*'s Northern Expedition. By daylight, Japanese planes from an aircraft carrier stationed offshore were bombing Chinese positions and heavy fighting had broken out in many sections of the city.

Unhappily, Japanese bombs also hit civilian targets, including a Chapei railway station that caught fire. A Western journalist photographed a small Chinese baby, burned, crying in the ruins. It made the front pages of American newspapers the next day. That swiftly, the Shanghai incident became an international crisis.

The American Secretary of State, Henry Stimson, had followed the fighting in Manchuria and Shanghai with increasing dismay. He had hoped that Shidehara would restrain the military's hardliners. Shidehara's fall and the Chapei bombing convinced him that Japan was out of control and had to be restrained. He sounded out the League's members, especially the British, and even considered calling a conference under the Nine-Power Treaty signed at Washington a decade earlier. But the British and French were cautious. The Manchurian issue was complex, they argued, and it was better to wait for the League's commission to report before taking action there. They agreed that Japan's actions in Shanghai were distressing, but the Chinese had offered provocation. Besides, all the economies of Europe were troubled and some radical leaders, such as Mussolini in Italy and Hitler in Germany, were emerging. It was no time for a major commitment to East Asia. An American approach to the Soviet Union was out of the question, given its leaders' communist ideology. In a mix of frustration and determination, Stimson proceeded on his own. In mid-February he issued a public letter implying that Japan's apparent violation of the Nine-Power Treaty's pledge of non-interference in China's domestic affairs might make all the other Washington treaties null and void. His reference was obvious: the battleship treaty. Stimson was obliquely threatening Japan with a renewal of a naval arms race if Tokyo refused to call its military to heel.

Between America and the Imperial Army, Prime Minister Inukai was in an impossible position. Stimson had called for a peaceful resolution of the Shanghai incident, and his reference to the Nine-Power Treaty clearly signalled America's opposition to Japan's resolution of the Manchurian question. Yet to reverse that resolution was absolutely inconceivable. The army, the public, and most of Inukai's own *Seiyūkai* were adamant on this point. By the same token, Inukai could painfully recall the reason for his appointment as premier: Saionji, and by implication the Emperor, wanted the military restrained.

Inukai tried to walk the tightrope between these conflicting pressures. He had to protect Japanese lives and property in Shanghai, so he approved the dispatch of reinforcements to Shanghai, where the marines, however skilled, found themselves badly outnumbered by Chinese forces. The army's fresh troops pushed the *Kuomintang* armies out of Shanghai, but broke off pursuit just before the League of Nations reconvened in early March. The ploy did not entirely work, as the League successfully insisted on appointing a truce committee that would oversee the withdrawal of Japan's forces from the area in exchange for China's agreement to safeguard the Japanese residents in Shanghai. The League would have preferred a committee with the power to set a deadline for the pull-out, but an outraged Imperial Army considered this a violation of its right of supreme command and threatened to have Japan withdraw from the League rather than comply.

In this way, Inukai got through the Shanghai incident. But Manchuria promised to be far more difficult. Japan's Kwantung Army there had made no secret of its wishes. Back in October 1931 it had publicly called for a new government for Manchuria – really a puppet government. Uchida Yasuya, one of Shidehara's right-hand men who had been named President of the South Manchurian Railway, turned on his mentor and declared his support for an 'independent' Manchuria. By early 1932, with Shidehara out of office and Chang Hsüeh-liang's forces out of Manchuria, the matter was decided. On 1 March 1932, the new 'nation' of Manchukuo was established in defiance of the League's wishes and Stimson's implicit threats.

Inukai could do nothing about Manchukuo's creation. But he did withhold Japan's extending diplomatic recognition to its infant regime. Aware that the Shanghai incident had badly damaged Japan's relations with the West, especially the United States, Inukai was determined not to provoke a final break by such a provocative action over 'Manchukuo'.

It was an empty and ultimately counter-productive gesture. The West was unimpressed; Inukai had done nothing to address China's loss of territory. And the Prime Minister's delay in completely bowing to the wishes of the military led to his assassination on 15 May 1932 at the hands of young naval officers.

Inukai's death compounded Saionji's difficulties in selecting Japan's next premier. The *Seiyūkai* had a commanding majority in the Diet, but the party's surviving leaders all strongly opposed the London Naval Treaty and favoured close collaboration with the army. The *Minseitō* had few seats in the Diet and no strong leaders except Shidehara, who was anathema to too many other Japanese élites to ever hope to serve as Prime Minister and who probably would have been assassinated anyway. Saionji considered providing the *Minseitō* with a *de facto* leader by naming either a retired admiral or General Ugaki as Prime Minister to rein in military hotheads. But, for precisely that reason, neither could become premier. The navy refused to allow any former naval minister to serve; the army blocked Ugaki's candidacy. No one who favoured a moderate foreign policy could hope to be Prime Minister of Japan in 1932.

Saionji refused to allow any of the militaristic *Seiyūkai* leaders to become Prime Minister, so he picked the bland Admiral Saitō Makoto. Both political parties joined Saitō's 'national unity' cabinet, but the top prize – the foreign minister portfolio – went to the army's new favourite: Uchida Yasuya.

Uchida's selection was a substantial setback to Saionji's hopes for moderation. Uchida came to power just as the League of Nations' commission had arrived in Japan to complete its investigation of the Manchurian incident. The new Foreign Minister, fresh from duty as President of the South Manchurian Railway, had very strong pro-army views. He also used very strong words with the commission. Manchukuo existed and could not be ignored, he asserted. Japan would not cooperate under the Nine-Power Treaty in determining its relations with the new state, nor would it heed any recommendations of the investigative commission. Before the Diet, Uchida averred that Japan would defy the entire world over Manchukuo if necessary. By mid-September, that is exactly what Japan did by extending diplomatic recognition to the puppet state.

A few weeks later the League's commission published its report. It admitted that Manchuria had been a troubled and turbulent land and that Japan had legitimate and recognized rights there. But the creation of Manchukuo was artificial and contrary to China's territorial integrity. The report recommended negotiations between China and Japan that would lead to stability, a recognition of Japanese rights, and restoration of Chinese sovereignty.

To reply, Uchida sent Matsuoka Yōsuke, who also had been a senior official of the South Manchurian Railway, to the League. Matsuoka immediately made clear that Japan rejected the report's recommendations completely. In February 1933, after the League endorsed them anyway, in a vote of forty-two nations against one, Matsuoka led Japan's delegation out of the League.

Japan was determined to go it alone. By the end of 1934, its government had given the required two-years' notice that it would no longer be bound by the naval treaties negotiated at Washington and London. And its Foreign Ministry had declared that no other nation or organization of nations had any jurisdiction over East Asia except through Japan. Freed from international obligations and constraints, Tokyo hoped to create a new order in East Asia.

Within Japan there were hopes for a new order, too. Army officers who had built a centralized, planned economy in Manchukuo wanted to do the same for Japan. They wanted to do away with the stale ideas of political parties competing for power and private corporations competing for money. If all Japan could be united in a common goal, Japan could unite all East Asia. The resulting bloc – the resulting Japanese Empire – could withstand any challenge, military or economic. Then, and only then, would the final goals of the Meiji Restoration be realized. To these officers, this was the path to true independence and security. It turned out to be the road to disaster.

Notes

[1] Older warships would be scrapped instead.

[2] In addition to the three major naval powers – Britain, Japan, and the United States – France and Italy also signed the agreement.

[3] In this respect, it closely followed the terms of the Lansing–Ishii agreement of 1917.

[4] Even upon its founding in 1906, the South Manchurian Railway was the largest company in Japan's experience. It came to dominate the economic development of southern Manchuria in all respects: mining, agricultural, and industrial. Manchurian development was substantial and rapid. The population of the region mushroomed to over 30 million in 1930, well over double the 1900 figure.

[5] Yamamoto's given name is sometimes rendered Gombei.

[6] The Kwantung Army was a small Japanese force stationed on southern Manchuria's Kwantung peninsula since the end of the Russo-Japanese War in 1905. Its chief task was to protect the Japanese-owned South Manchurian Railway.

[7] The officer's name was Colonel Kōmoto Daisaku.

[8] The *Seiyūkai* was now under the control of Inukai Tsuyoshi. Tanaka had died in late 1929.

[9] Cited in Kobayashi, T. 1984: The London Naval Treaty, 1930. In Morley, J.W. (ed.), *Japan erupts: The London Naval Conference and Manchurian Incident, 1928–1932*. New York, NY: Columbia University Press, 72.

[10] The term 'Shōwa' came from the name each emperor chose to designate his rule when he came to power. Thus one might speak of Meiji Japan when referring to Japan from 1868 to 1912, Taishō Japan from 1912 to 1925, and Shōwa Japan from 1935 to 1989. Japan is currently in the Heisei era.

6
New order against old

Japan struck out on its own by leaving the League of Nations in 1933 in defiance of the old order of international relations. At the same time, its leaders, mainly its army, attempted to overthrow the old order within Japan, too. Japan's new order, therefore, would have two parts. Abroad, Japan would pursue an independent course, casting aside all hope of cooperating with the West. Manchuria, now the puppet state of Manchukuo, would be an integral part of the Japanese Empire. The rest of China, especially the five northern provinces bordering Manchuria, would fall under Japan's influence, though they would not be formally annexed. The Soviet Union, itself not a part of the West's old order, would be neutralized by Japan's association with Germany, newly resurgent under Adolf Hitler and his National Socialist Party. Japan would use its new acquisitions to build up a power base so great that no Western power would be able to challenge it in the future. Japan, in short, would at last be secure through the construction of its own vast empire.

This ambitious international programme also required a new order within Japan. In part, this domestic reform was necessary to rid Japan of leaders who opposed the new foreign policy. These leaders included members of the business community who believed that trade with China and the West, not confrontation and isolation, was the best course for Japan. There were also politicians and even members of the Emperor's Court (and quite likely the Emperor himself) who agreed with these business leaders. In part, as well, domestic reform was necessary because Japan's existing political and economic structures were not well geared to going it alone. It would do little good to exploit coal from Manchuria if that coal simply went to power a Japanese textile mill that sold cloth to America, for example. Acquiring the provinces of northern China would be useless if the resources there simply made Japanese businesses richer. Instead, reformers, usually coming from the army, argued that the entire Japanese state had to be regeared for total defence. Naturally, the army would supervise the regearing, and would control the new Japanese 'National Defence State'.

The story of Japan's foreign relations from 1933 to 1945, therefore, has two themes. The better known of these is Japan's pattern of expansion abroad and, of course, the end of that expansion at the hands of the Americans beginning in 1942. Just as important, though, were the reformers' attempts to reorder Japan from within. Ironically, they were much more successful in overcoming foreign resistance than in ousting their domestic opponents from power. The result, by 1945, was a Japan surprisingly receptive to a swift return to cooperation with the West the instant the Pacific War was over.

But in the early 1930s, the army had the upper hand. Its first triumph was to reorganize the way the Japanese Cabinet did business. Instead of the full cabinet discussing policy decisions, all important matters were referred to an 'inner' cabinet of only five ministers: Army, Navy, Foreign, Finance and the Prime Minister, a ratio of military to civilian ministers much more to the army's liking. The army also stripped the Foreign Ministry of its control of relations with Manchukuo – technically an independent nation – with the creation of an army-dominated Manchurian Affairs Bureau in 1934. At the same time, the economic activities of the *zaibatsu* in Manchuria and even the South Manchurian Railway were increasingly brought under strict army control. In many respects, Manchukuo was to be the experiment for, and forerunner of, a reordered Japan.

The political parties likewise suffered a loss of importance in the early 1930s. The selection of Admiral Saitō Makoto as Prime Minister in 1932 had been a bitter blow to both the resurgent *Seiyūkai* and outgoing *Minseitō*. The 'inner' cabinet mechanism made it all but impossible for party leaders to influence important policy decisions. Predictably, members of both parties fell into squabbling over how to respond. Some leaders in both parties felt it was better to join the army and embrace its foreign policy of expansion and domestic reforms. Others were quite willing to go along with a hardline foreign policy, but feared that the army intended to destroy the political parties along with the rest of the old order inside Japan. They opposed too close a connection with the military and resisted its domestic policy initiatives. A very few felt misgivings about the army's foreign as well as domestic policies. When they spoke up, they often were assassinated by young army officers.

The Imperial Army, therefore, had very little resistance to its foreign policy programme from the civilian élites. The Imperial Navy, however, was another matter. Once anti-Western hardliners such as Katō Kanji and Suetsugu Nobumasa took control of the navy, they were willing to permit the army its conquests in China. But, in return, they wanted the army's support for their desire to break the London treaty and increase the size of the fleet. The first result of this mutual back-scratching was a formal decision on the fundamentals of Japan's foreign policy. Drawn up in October 1933, this document declared that there would be twin crises by 1936. In that year, Japan would refuse to renew the London Naval Treaty. At the same time, the army would have completed a three-year programme to exploit its holdings in Manchuria to prepare for war with the Soviet Union.

Under the terms of this agreement, the combined spending of the two services would rise from one-third to nearly one-half of Japan's total budget. Finance Minister Takahashi Korekiyo was unhappy over that prospect, but he was overruled by the army–navy steamroller.

Interestingly, another group also opposed this 1933 accord with more success. A small but influential group of army officers themselves thought that it was premature to prepare for war against the Soviets so quickly. They argued that Japan needed much longer than three years to complete its new order. It was better to develop Manchuria and to bring northern China under firm Japanese control before facing off against Moscow. These officers also urged fundamental reforms within Japan before any major conflict could be contemplated. In a showdown within the army, they succeeded in seizing power, making one of their own, General Hayashi Senjurō, the new Army Minister. Hayashi's rise ensured that, going into 1934, the army would press hard for further acquisitions in China and changes inside Japan while downplaying the possibility of war against the Soviet Union. Hardliners in the Foreign Ministry, old opponents of Shidehara Kijūrō, were delighted by this turn of events and had one of their members, Amō Eiji, newly appointed as head of the ministry's information services, announce an 'Amau Doctrine'[1]. It asserted that Japan alone was responsible for maintaining peace and order in East Asia, much as the United States had done in the Western Hemisphere. Attempts by other powers to interfere, such as providing assistance to China, would not be tolerated.

Amō was not speaking on his own. On the contrary, his announcement indicated how close the Foreign Ministry had moved toward the expansionist position. The new Foreign Minister, Hirota Kōki, had endorsed the doctrine in instructions to Japanese diplomats in China and before the Diet shortly after taking office in late 1933. Unsurprisingly, Hirota raised no objections to the army's creation of the Manchurian Affairs Bureau. Nor did he oppose an inner cabinet decision to proceed with further acquisitions in China in 1935.

Consequently, the army had a clear field in its move on the five northern provinces of China that year. As a result of the truce reached between Japanese and Chinese forces after the Manchurian incident, the *Kuomintang* army had agreed to evacuate a portion of these provinces already. Moreover, the Japanese still had rights to station small garrisons

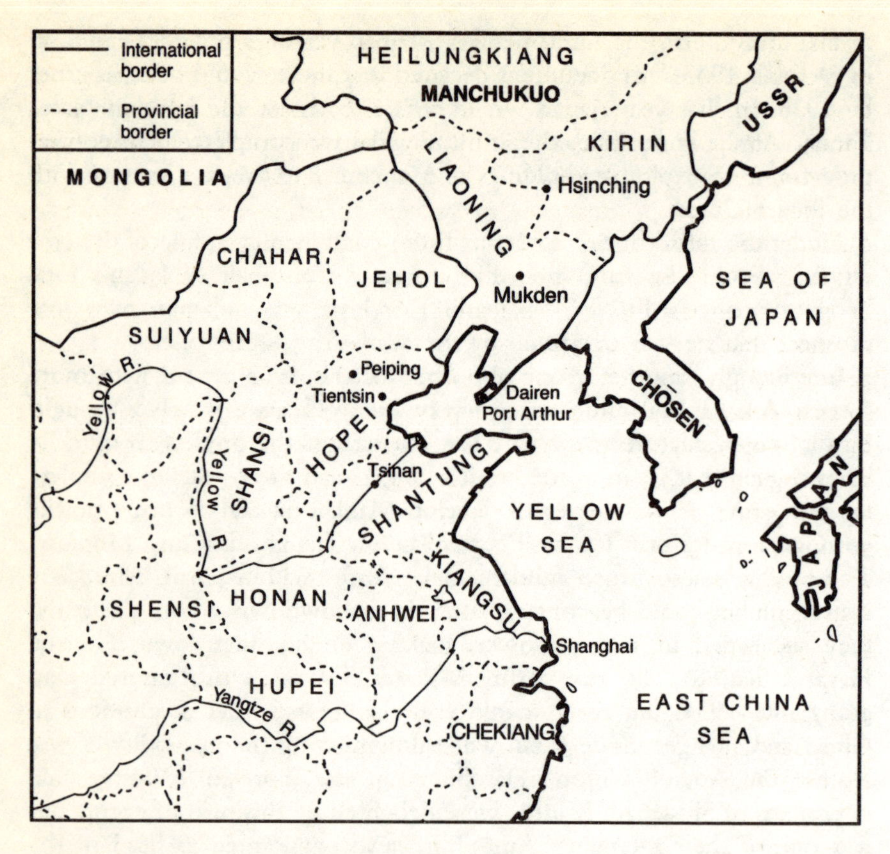

Map 6 Manchuria and northern China 1935

around Peking and Tientsin dating from the Boxer rising of 1900. In the spring of 1935, the Imperial Army used the murder of two pro-Japanese Chinese newspapermen as a pretext to demand the removal of the Nationalist forces from two of the five northern provinces.

Kuomintang leader Chiang Kai-shek did not feel confident enough to risk a military confrontation with the Imperial Army. Instead, he resorted to financial tools to keep northern China out of Japanese hands. With British help, he began to centralize control of all Chinese currency under banks in his control. The Japanese army sought to prevent the use of his Nationalist currency in the five northern provinces, but without much success. Instead, the army had to settle for the creation of an 'autonomous' regime in two of the provinces that was a pale imitation of the true independence from Chiang's authority achieved for Manchuria. Somewhat frustrated, army leaders resolved to do better in 1936.

The army's efforts at reform within Japan went better, at least at first. In the summer of 1934 the Cabinet Investigative Bureau was established. It was a vehicle for reformist army officers to meet and ally with like-minded bureaucrats from the various civilian ministries. Reformers from the Ministry of Agriculture and Forestry, for example, were pleased with the idea of a strong government hand in directing the rural economy. Many younger members of the Ministry of Commerce and Industry objected to the *zaibatsu*'s control of industrial cartels established to alleviate the impact of the Great Depression. With the army's help, they hoped to make those cartels real tools of the state, not private business. Never before had the army's political influence been spread so widely through Japan's political leadership. Only very conservative ministries, such as Finance, opposed the army's ideas. As a result, the bureau actually drew up specific plans to reform Japan along the army's lines. A central office for budgeting was to be created directly under the Prime Minister – a proposal aimed directly at the Finance Ministry's ability to restrict military spending. A personnel agency would be established to isolate the bureaucrats from the influence of the political parties, a favourite idea of the army going back to Yamagata's time. Most civilian ministries would be merged and reduced, their powers diverted to army-run 'mobilization agencies' in charge of food, fuel, and other necessities of any war effort, especially energy. The army especially hoped to bring all of Japan's electric utilities under direct state (meaning army) control. The capstone of the army's efforts was a colossal five-year plan to expand Japan's ability to produce the sinews of war, from steel to (synthetic) petroleum. Drafted by Ishiwara Kanji, a guiding force during the Manchurian incident, the plan called for immense investments in these heavy industries.

How was Japan to afford these investments? For Hayashi and his allies in the army, the answer was clear. Spending on civilian programmes had to be minimized. Even regular military budgets had to be kept low. Because Japan would rely heavily on imported goods to build the steel mills and petroleum plants called for in Ishiwara's plan, it was also imperative not to force a break in Japanese relations with the West, especially America and Britain.

This need for a long peace with the West while Japan built up a war machine at home made sense, but it ran contrary to Japan's equal need to gather resources for that machine by conquering China and perhaps other parts of Asia. This was a dilemma for Japanese foreign policy that the Imperial Army's reformers never resolved.

They also had difficulty, predictably, with other élites in Japan. Finance Minister Takahashi was aghast at the expense of the army's plans to remake the entire Japanese economy. The navy suspected (rightly) that the army was plotting to deprive it of the funds and materials to commence a large naval build-up as soon as the London treaty was broken. Even within the army, advocates of swift expansion at China's expense felt that the programme's insistence on stable relations with the West would rule out provocative steps and slow their progress on the continent – as indeed it did, or at least should have done.

The political parties' reactions were more complex. In the aftermath of the Manchurian crisis, both parties grudgingly had gone along with Admiral Saitō Makoto's appointment as Prime Minister. But when he retired in 1934 and the parties were again denied the premiership in favour of Okada Keisuke, party members became rebellious. In the *Seiyūkai*, a three-way split developed. Some openly embraced the army's domestic reforms and advocated expansion abroad even more vigorously than the army's leaders. These politicians often spoke for (or even worked for) new industrial companies which stood to gain immensely if the army's plans went forward.[2] A second group was made up of opportunists who joined the Okada Cabinet in hopes of power and favour.[3] They were read out of the party for their move. The rest of the *Seiyūkai* followed Hatoyama Ichirō, a traditional conservative who shared the anti-radicalism of the army and its allies, but felt that the army itself was slipping into radicalism with its sweeping economic reform plans. Hatoyama and the traditionalists also wanted to preserve the power of the regular political parties by demonstrating that the Diet's cooperation was still important.

The *Minseitō* was likewise riven. A minority felt that Hatoyama was right and that the army's challenge to the old order was so great that alliance with the *Seiyūkai* was the only choice left. But most members preferred a separate stand against the army. That meant hunting for a prestigious leader for the party to replace the slain Hamaguchi, a hunt which presented its own difficulties. One possibility, General Ugaki Kazunari, had great prestige and stood for a foreign policy of moderation. For exactly that reason, the army's current leadership detested him. Ugaki, fully aware, insisted upon a merger with the *Seiyūkai* in order to lead the combined parties against the army and its allies. That condition made him unacceptable to most of the *Minseitō*. Prince Konoe Fumimaro certainly was prestigious, but he was drawn to the idea of a Japan reformed in many respects along the army's lines, with himself leading a new, mass political party in the style of Germany's National Socialists. This idea repelled the traditionalists in the *Minseitō*.

In some respects, Hatoyama indeed was right. Although the parties could not regain the premiership, they still controlled the Diet. There, they blocked the army's plans for reform year after year. By early 1936, the army's leadership had resolved upon sterner measures to break the deadlock. The generals hoped to enlist a prestigious political figure themselves, use him to create a new political party, and vote the old *Seiyūkai* and *Minseitō* out of the Diet. Konoe was the obvious candidate, with his reformist ideas. But the prince wanted time to think over the possibilities.

These delays were too much for younger army officers, who were impatient for their 'Shōwa Restoration' at once. On the morning of 26 February 1936, 1400 army troops under the command of these junior officers seized government buildings in Tokyo and hunted down high governmental officials who had obstructed the army's reforms. They assassinated former premier Saitō, Finance Minister Takahashi, and General Watanabe Jōtarō, the conservative inspector general of military education. Prime Minister Okada escaped only because the rebels mistakenly murdered his brother-in-law instead.

These revolutionaries, led by very junior officers, not the army's top men, acted out of a variety of motives. In essence, they thought that Japan's leaders were uncaring, corrupt, and unable to address the foreign menace. They were convinced that a Western-style capitalistic economy was inherently greedy, but, much more than that, they thought that capitalism was a failed and dying system. For this reason, they wanted to go much further than even the senior army reformers like Ishiwara and Hayashi. Japan did not have time to build steel mills and invest in other heavy industries, according to the young rebels. Doing so only deepened the unholy alliance between the moguls of business and the army, which ought to remain true to the Emperor and spirit of Japan alone. The junior officers insisted upon the wholesale removal of the old ruling élites. Only the army was pure enough to be entrusted with the Emperor's country. These officers saw themselves ousting an old order much as the Meiji Restoration had done seventy years earlier.

It was a poor analogy. The Meiji leaders had carefully cultivated many of the old élites, assuring them that their place would be secure under the new regime. The February rebels issued proclamations openly denouncing all of the old leaders. Meiji had limited resort to violence; the February plotters embraced murder from the outset. Most importantly, the Meiji oligarchs relied upon the Emperor to legitimize their rule. In February 1936, no single person was more responsible for crushing the rebellion than Hirohito. In one of his very rare departures from passive observation and consent, the Emperor demanded swift punishment of the rebel forces. Loyal army troops brought the would-be insurrection to a close within two days.

Although short-lived, the February rebellion deeply affected Japan's politics and foreign policy. Prime Minister Okada resigned after his brush with fate. His successor, Hirota, was much more amenable to the army's domestic reforms and expansionist foreign policy. Needless to say, the army's reform-minded senior officers enjoyed unquestioned control over the army after the rebel leaders had been executed and their sympathizers removed from the service.

Although Hirota was Prime Minister in name, the real power belonged to Army Minister Terauchi Hisa'ichi, who determined the cabinet's make-up. With control of the cabinet well in hand, Terauchi and the army moved to make their domestic reforms a reality. Proposals to nationalize Japan's electric power industry and to create a Health Ministry (to ensure a reliable supply of vigorous men to be soldiers), among others, were readied for the Diet's approval. As well, work progressed to bring Ishiwara's colossal production expansion plan into reality.

Hirota went along with these proposals, but he was poorly prepared to support them before the Diet because he was a career diplomat without any party affiliation. His Finance Minister, Baba Eiichi, raised no budgetary objections to the army's plans, as his murdered predecessor Takahashi surely would have. But even Baba was worried about Japan's ability to pay for the army's grand plans at the same time it undertook a terrific naval construction programme.

In fact, the biggest problem for the army after the abortive coup turned out to be the navy. For the Imperial Navy, 1936 was the year of deliverance. The despised naval treaties finally expired, leaving the navy to build warships free of constraints from abroad. Constraints within Japan were another matter. The army's foreign policy called for good relations with Britain and America. Imports from both were necessary for the army's production expansion plan. But a massive naval construction programme would antagonize these Western powers. At the same time, every ton of steel used to build battleships was a ton that could not be used to build more steel mills or synthetic petroleum plants. For this reason, the army opposed the navy's expansion.

Yet the army was reluctant to openly sabotage the navy's building plans. In part, the army feared that the navy would seek revenge against the army's far larger spending proposals. In part, the army wanted the navy to agree to an alliance with Nazi Germany.

The German alliance was a centrepiece of Japan's foreign relations for the army. It would end Japan's diplomatic isolation, so apparent since the end of the Manchurian crisis, and make it more difficult for other nations to attempt to resist the army's further encroachments in China. And it would provide a partner against the Soviet Union, which had increased its forces along the Manchurian border with startling speed. In fact, the chief benefit of a Japanese–German alliance would be to intimidate Moscow.

The navy – and Hirota – saw an alliance with Germany in a different light. A vague agreement could usefully serve as a warning to Moscow not to interfere with Japan's peaceful penetration of China. It could also help persuade the British, who still held a large economic stake of southern China, to agree to a larger Japanese role there. But an actual agreement to go to war with Germany against the Soviets was out of the question under any circumstances. The navy was especially concerned about the matter of geography in China. It was eager to have Japan's formal foreign-policy statements recognize that China's southern areas – of naval concern – were equally as important as the northern ones of the army. Put another way, the navy wanted its rationale for a larger fleet recognized so that it could build that larger fleet.

The result of this inter-service wrangling was the 'Fundamentals of National Policy', drawn up in August 1936 and an army–navy agreement on force levels drawn up two months earlier. The army got its agreement with Germany, but in the watered-down form preferred by the navy. In September, Japan concluded an 'Anti-Comintern Pact' with Germany that fell far short of a military alliance. The two powers merely agreed to exchange information on the Soviet Union's agency for international communism, the Comintern, and not to aid the Soviets if either were attacked. The navy agreed to support the army's production expansion plan, but only if it was able to proceed with its immense warship construction plan for a fleet of a dozen battleships and a like number of large aircraft carriers with over a hundred lighter supporting ships and a massive fleet air arm. Moreover, the navy received the right to rule Taiwan directly (just as the army earlier had secured the right to govern Korea and Manchuria) and funds to sponsor the economic development of the areas further south, again in mimicry of the army's activities in Manchuria and north China.

The army was unhappy with the navy's new, aggressive stance, but could console itself that its core programmes, a connection with Germany and domestic reforms such as the production expansion plan, were official cabinet policy. Unfortunately, all these needed the Diet's approval, and the legislature's opening session in December 1936 saw both major parties in inflexible opposition. Indeed, in January, one *Seiyūkai* Diet member openly accused Army Minister Terauchi of plotting tyranny and Ozaki Yukio, an old Diet veteran of the 1913 constitutional crisis with the military, blasted army officers for their arrogance.

With a powerful prime minister allied with them, the army might have overcome this party opposition. But the army had agreed to Hirota's appointment as premier precisely because he was not powerful. Faced with the prospect of having none of its programmes adopted, the army decided Hirota had to go. Terauchi's resignation brought down the cabinet in February.

Ideally, the army would have preferred a prime minister who agreed completely with the army's programmes and who was strong enough to carry the Diet. Since no established party leader agreed with the army, it was necessary to look for a prestigious civilian who could form a new political party to sweep the old ones away. Once again, the obvious choice was Prince Konoe. As his title implies, Konoe came from one of the leading (and oldest) families in Japan. His father had been an influential noble in the House of Peers during the Meiji period. The son was now leader of the Peers and had made a name for himself by publishing writings critical of the Anglo-American world order in leading Japanese journals of opinion.

In short, Konoe was the army's ideal candidate to become Prime Minister. But Konoe still refused the army's invitation. He was all too aware of how badly the army's actions under Hirota had alienated Japan's other élites. Konoe counselled the army that it first had to win their cooperation by sharing power with them. That meant sharply moderating its reform programme, at least initially. A direct frontal assault on those other élites, by creating a new political party, for example, was likely to backfire.

Given a choice between watering down its reform programme under Konoe or fighting the old political parties directly, the army chose to fight. It vetoed the *Minseitō*'s bid for a cabinet under Ugaki. Instead, its own General Hayashi became Prime Minister. But Hayashi himself was something of a disappointment to army reformers, because he took Konoe's advice and began to moderate some of the army's programme.

This was not entirely by choice. Many army reformers viewed Hayashi's appointment as their best opportunity to achieve fundamental change. They sought to pack his cabinet with their radical allies. But every time, their choice was vetoed by Japan's political and economic leaders, who refused to cooperate with Hayashi's candidates. Conceding defeat, Hayashi reached out to these business and financial communities by accepting some of their nominees for powerful cabinet posts, such as Finance Minister,[4] and trimming his government's huge budget substantially. Moreover, Hayashi agreed to defer consideration of the army's key reforms, from nationalizing the electric power industry to commencing the production expansion plan.

In his foreign policy, too, Hayashi was forced to strike a moderate note. He appointed Satō Naotake as his Foreign Minister. Satō had been ambassador to France and a well-known advocate of better relations with the West. Moreover, he felt that the best way to achieve those relations was for Japan to cease encroaching upon China's territory and sovereignty. In this respect, he sounded remarkably like the despised and departed Shidehara.

Perhaps as surprisingly, Satō's views actually became Japan's policy. To be truthful, this was not entirely Satō's doing. Hayashi and Ishiwara understood that Japan could not jeopardize its chance for domestic reform by alienating the West over Chinese issues. As well, the Chinese had helped their own cause by the beginning of 1937 in two respects. First, Chinese forces had defeated a band of Mongols that the Imperial Army had supported in hopes of carving an independent Mongol regime out of Chinese territory. Second, after the Sian incident of December 1936, the Chinese leaders agreed to stop their internal feuding and meet the Japanese threat with a 'United Front' including even the Chinese communists.

Both foreign and domestic factors combined, therefore, to moderate Japan's stance toward China in early 1937. In April, the inner cabinet formally agreed to abandon attempts to create an independent Mongolia and to defer indefinitely making China's five northern provinces autonomous. To avoid any repetition of 1931, Tokyo specifically ordered its army units in China to take no provocative action. Instead, the Hayashi Cabinet would stand pat in China while patiently building a political coalition in Japan that would see eventual adoption of the army's domestic reforms.

As it turned out, patience was in short supply in both China and Japan. Hayashi's concessions to other élites and his consequent slow pace of reform irritated many in the army without paying off in better relations with the political parties. Hayashi offered cabinet posts to established members of the *Seiyūkai* and *Minseitō*, but stipulated that they would have to resign their party memberships in order to accept the posts. No one accepted. Once the Diet session ended, Hayashi hastily called for national elections in April. Hayashi and the army hoped that many reformists would win seats. The results were a bitter rebuke to them, as the two major parties maintained their domination of the house. Unable to form a cabinet that would have any hope of seeing the army's reform programme pass, except in greatly weakened form, Hayashi resigned.

The new Prime Minister was Konoe. In many respects, he was the only possible choice, acceptable to the parties, the business community, and the disgruntled military. Konoe promised to mediate among these élites, and his cabinet appointments showed it. Hirota returned as Foreign Minister. He was pledged to working for better relations with the West while doing nothing to undermine the army's accomplishments against China. The *Minseitō's* representative[5] strongly favoured nationalizing electric power and other army reforms. The *Seiyūkai* sent Nakajima Chikuhei, who headed a company rich with army contracts for airplanes. Although these men were mavericks, the parties' established leaders were pleased with Konoe's appointment, since he still refused to lead a new political party against them. In sum, Konoe came to power in June 1937 hoping to restore political consensus to Japan while moving forward with a slow, moderate version of the army's reform programme at home and a cautious, non-provocative foreign policy.

Konoe did restore that consensus, but only at the price of a foreign policy that would overwhelm him and, in time, the Japanese Empire itself. That price would be paid in China.

From Tokyo's point of view, Japan had been a model of restraint in China during the first half of 1937. No effort had been made to build on the pro-Japanese autonomy movements in the northern provinces since 1935. In fact, to guard against the Kwantung Army acting on its own, as it had done in 1931 in Manchuria, the army's central headquarters had removed northern China from the Kwantung Army's area of responsibility. A separate command, based in the Chinese cities of Tientsin and Peking, had been established.

But Chiang Kai-shek, and Chinese leaders directly in the northern provinces, interpreted these moves as a prelude to another 1931. Japan, in their eyes, had abandoned working with local Chinese collaborators in north China just as it had abandoned Chang in Manchuria before the Mukden incident. The creation of Japan's separate northern China command seemed ominous, not reassuring. Moreover, Chiang was under intense pressure to stand up to the Japanese this time, as he had not done 6 years earlier. Patriotic Chinese had gone so far as to abduct him in late 1936 – the famous Sian affair – and had refused to release him until he promised to show more backbone.

Whatever one's perspective, the fact remains that Japanese troops, in however small numbers, were on Chinese soil around Tientsin and Peking. The possibility for a blow-up was ever present. That is exactly what happened on 7 July 1937. Japanese and Chinese troops exchanged fire (each claimed the other shot first) near the Marco Polo Bridge just outside Peking.

Reactions among Japanese policy-makers split three ways. Those army officers chiefly concerned with their reform programmes aiming to achieve Japan's long-term goal of economic self-sufficiency argued for a compromise with the Chinese to settle the incident immediately. Nothing could be allowed to divert funds and resources from the production expansion plan that, under Konoe, seemed likely at last to win the Diet's approval. Moreover, hostilities with China would lead inevitably to friction with the West, especially the United States – key supplier of many of the goods that Japan's new order required for its realization. Ironically, in July 1937 these officers were led by Ishiwara Kanji, who, 6 years before, had defied Tokyo's caution and had proceeded with the conquest of Manchuria, dragging his superiors along.

Most Japanese field officers in China argued that Ishiwara had set the right example in 1931 and had the wrong arguments in 1937. These hardliners maintained that the Chinese would not resist forceful Japanese moves in north China any more than they had in Manchuria earlier. The correct response to the fighting around the Marco Polo Bridge, therefore, was to threaten harsh measures and demand concessions from Chiang, perhaps the long-desired Japanese dominion over the five northern provinces. These officers doubted that Japan would have to back up its threats. They were certain that the Chinese would back down.

A third group of Japanese leaders, typified by Konoe, was essentially opportunistic. While the Prime Minister certainly wanted to avoid a long, draining war, a short, victorious conflict would be ideal for securing the Diet's approval of a wide array of reform programmes. Patriotic fervour would ensure their passage. Once the shooting stopped, Konoe would have his new order secure at home and abroad with victories in the Diet and over Chiang. In sum, Konoe saw little risk in the hardliners' position and a great deal of gain.

Of course, Konoe as Prime Minister had no power to order the army to either settle or escalate the 'Marco Polo Bridge Incident'. That power was held by Ishiwara, who occupied the key position of head of the Army General Staff's Operations Division. For several days, he used his post to block reinforcements for the Japanese forces in north China. But he succumbed to his opponents' whipsaw arguments that, on the one hand, even preliminary mobilization of Japanese reinforcements would cow Chiang into submission and that, on the other, if the Chinese somehow did elect to fight, Japan's small garrison in north China would be quickly overwhelmed by sheer numbers and its massacre would be on Ishiwara's hands. The hardliners also assured Konoe that a minimal dispatch of force, if one were needed at all, would be sufficient to secure north China within several months, if not sooner. So Ishiwara capitulated and Japan sent a small army to north China.

Ishiwara may have surrendered to the Imperial Army's hardline field officers, but the Chinese did not. They fought stubbornly in the northern provinces and, to the Japanese army's amazement, dramatically widened the conflict by attacking Japanese possessions in Shanghai, far to the south. For Chiang, the Shanghai assault was a masterstroke. It caught Japan off guard and, perhaps more importantly, ensured the involvement of the Western powers, which themselves had important holdings in that large port city.

Although Chiang could not have known, he certainly would have been delighted by the effects his Shanghai attack achieved within Tokyo. The Imperial Navy had been suspicious that the army and Konoe were conniving to use the Marco Polo Bridge Incident to pass army reforms and enhance the army's power. Naval leaders had consented to sending a small army to north China only reluctantly and only after securing the army's promise that if trouble spread southward in China, into the navy's area of responsibility, army assistance would be forthcoming at once and without stint. The army hardliners had made the pledge easily, confident that the Chinese would never escalate hostilities. Chiang's action now compelled them to make good their word – and radically expand what was renamed the 'China Incident' in the process.

Even so, the hardliners used the same assumptions to govern their response in Shanghai: shown enough Japanese force, the Chinese would yield. But yield what? After weeks of fierce fighting, the Japanese were able to push Chinese forces out of Shanghai. In the north, Peking and Tientsin were secured. But Chiang refused to recognize Japanese control over the five northern provinces and refused to end the conflict.

To dispose of Chiang swiftly, the army drove on his capital city of Nanking and created a rival (Japanese-puppet) regime to replace him. Nanking fell in mid-December, but the results were bitterly disappointing. Japan already had alienated the West and the Soviet Union by continuing the war against China. A conference of Western powers, held in October at the Belgian capital of Brussels under the provisions of the Nine-Power Treaty, had not led to any concrete steps against Tokyo, but had shown publicly growing Western dissatisfaction with Japan. That dissatisfaction turned to outrage when, in occupying Nanking, Japanese forces hideously murdered huge numbers of Chinese in the city, perhaps as many as 200,000 in what came to be called the 'Rape of Nanking'. And Japanese aircraft attacked and sank the United States gunboat *Panay*, killing three American sailors.

Japan swiftly moved to resolve the crisis with America, offering apologies and indemnities to the bereaved American families. But its treatment of the Chinese was quite different. In mid-January 1938, Konoe declared that Japan no longer recognized Chiang Kai-shek as the legitimate ruler of China. A new Chinese regime, yet to be finalized by its Japanese masters, would govern instead.

Konoe's statement arose from extreme confidence, even arrogance, in Japan's ability to win the China conflict quickly and decisively. In retrospect, it appears exceptionally foolish. Chiang was unbowed. He still commanded substantial forces of his own. More to the point, rival Chinese leaders, such as the communists, were hardly about to collaborate with Japan. They much preferred to maintain a united front with Chiang against Tokyo. That meant that there were no real élites in China that Japan could use to create its rival regime. As importantly, though no one in Tokyo then knew, secret naval conversations for possible joint action against Japan were begun between Great Britain and the United

States. Moreover, the Soviet Union had begun military aid, modest but symbolically and psychologically important, to Chiang.

That Soviet aid was not secret. By early 1938 it disturbed a small but important number of officers in the army's General Staff, who felt that Japan had already committed far too many troops to the China war (over half those available, in fact) at the expense of preparations against the Soviets. They pressed for negotiations with Chiang. But Konoe's speech made talks nearly impossible.

In fairness, Konoe and the army hardliners had some reason for their optimism. Chiang had troops, to be sure, but his best had been expended in the battles around Shanghai at the start of the conflict. He had fled Nanking to set up a new capital in Hankow; it could be easily seized in 1938 to demonstrate Chiang's impotence and irrelevance. It was regrettable that no popular Chinese leaders had been found who would collaborate. But the Imperial Army had operated Manchuria for six years without notable collaborators there. True, the Soviets were assisting China. On the other hand, Germany aligned itself with Japan in February, even promising diplomatic recognition of Manchukuo.

This last point merits emphasis. Under the Nazis, Germany had made a substantial effort to assist Chiang Kai-shek's Chinese Nationalists militarily. By publicly supporting Japan, Berlin discarded years of work in China. But Japan was worth more to Berlin as a counterpoise to both the Soviet Union and the Western democracies by 1938. In that year, Germany hoped to score gains in central Europe, especially Austria and Czechoslovakia.

The Imperial Army, for its part, valued closeness with Germany for parallel reasons. Germany's power could intimidate the West and the Soviets. As importantly, Germany's government represented the sort of new order that the army hoped to create within Japan.

Early 1938 saw significant strides toward that creation. The keystone of reform was the National General Mobilization Law, which gave the government sweeping economic powers. For example, all industries could be compelled to form and join cartels to implement the government's mobilization efforts for war. The government could restrict access to capital for some industries and divert investment to other, more important ones. Individuals likewise could be drafted and placed at the government's whim, especially those with important professional and technical abilities.

The National General Mobilization Law was an important step towards the creation of the army's new order in Japan. But it was only a step, and a half-hearted one at that. Strong opposition from the established parties in the Diet compelled Konoe to agree that the provisions of the law would be invoked only during a declared war – not the current 'incident' in China. Konoe agreed, to the army's intense disappointment. But army leaders had only their own performance to blame for Konoe's reluctance to press for more. In early April, with the ink hardly dry on the mobilization law, the army had suffered a stinging setback at the hands of the Chinese. Japanese forces seeking to join the conquered northern provinces with Japan's latest gains around Shanghai and Nanking to the south were mauled in southern Shantung. An embarrassed army vowed revenge by moving on Chiang's latest capital of Hankow. But that was just the point: Chiang was still very much in the fight. Konoe, understanding as much, had no desire to confront Japan's political élites while Chiang's resistance persisted. At the same time, however, that prolonged resistance ensured that the terms of the mobilization law could not be invoked. The army's reform programme thus was stymied dually: while the fighting in China lasted, the mobilization law was inert, and Japan's resources themselves would go toward war in the present, not industrial production expansion for the future.

Of course, this was a nightmare for Ishiwara, Hayashi, and the other army reformers. They had warned all along that hostilities against China were wrong-headed. Even if China asked for a truce quickly, Japan's relations with the West would be badly damaged. If the fighting were prolonged, it would use up all the precious resources that should be diverted into building a heavy industrial base for Japan under the army's control. They had their reform law in the National General Mobilization measure, but it did them little good under the circumstances.

The key, as all understood, was to stop the fighting in China. To that end, Konoe and the hardline field officers of the army in China pursued a two-pronged and somewhat contradictory policy. The army, with still more reinforcements, seized Hsuchow city in the spring and made plans to drive on Hankow, Chiang's replacement capital, in the summer. Konoe pursued diplomacy. He changed two critical cabinet positions in May, making Ugaki Kazunari his new Foreign Minister and bringing in Itagaki Seishirō as Army Minister.

Konoe's shake-up was the move of a political master. During the debates over the mobilization law, angry party leaders had threatened to support a Ugaki cabinet to oust Konoe and his reformist impulses. By bringing Ugaki into his cabinet, Konoe deftly forestalled that possibility. But Ugaki was anathema to the army, since he had overseen a reduction in army forces in the 1920s and had argued for caution during the Manchurian crisis of 1931. To appease the army, Konoe made Itagaki Army Minister. Itagaki had been a leader of the Manchurian conspiracy and still favoured an aggressive policy toward China in 1937 and 1938, so he remained the darling of the army's hardliners. In sum, Itagaki was the price Konoe paid to get Ugaki into his cabinet.

It was a balancing act that worked well for Konoe inside Japan, but was a miserable failure in his attempts to resolve the Chinese situation. Ugaki strongly believed that the Chinese conflict was a disaster for Japan, despite its battlefield victories. The war virtually halted implementation of the production expansion plan. The fighting had compelled Japan to import huge quantities of war materials – from steel alloys to petroleum products to machine tools – and these came overwhelmingly from the United States, which was growing increasingly hostile to Japan as the China Incident continued. Indeed, some private American groups were already conducting public campaigns to halt exports to Japan, claiming that American scrap iron was being found in Chinese children by way of made-in-Japan bombs.

Ugaki, therefore, entered Konoe's cabinet only after securing the Prime Minister's assurance that the harsh statements of January were no longer operative and direct negotiations with Chiang were not only allowed but encouraged. Ugaki wasted no time putting out feelers to the Chinese, who were equally quick to respond.

In theory, Ugaki ought to have succeeded. His timing could not have been better. The Imperial Army had just taken Hsuchow and, as the Chinese knew, were preparing to drive on Hankow. The Chinese also knew that they could expect no help from the West, paralysed as it was by Nazi moves against Austria and Czechoslovakia. They should have been ready to sue for peace. Indeed, Ugaki was counting on the German distraction to induce the British to recognize Japan's position in Manchuria and north China.

That was just the problem. While Ugaki's diplomatic approach was a welcome departure from Konoe's early-year blustering, his words carried the same message as the Prime Minister's, only in a different tone. When Ugaki did begin his discussions with the Chinese, he could offer no concessions, except, of course, an end to the fighting. Both China and the West would have to recognize Manchuria as forever part of the Japanese Empire, north China was to be effectively removed from China's control, and Chiang himself would have to resign.

Whether the Chinese would have accepted these terms cannot be known, because they were not acceptable to Itagaki and the army. By the summer of 1938, the hardliners aimed at nothing less than a government for all China that would be, in essence, a Japanese puppet. Even as Ugaki carried on his conversations with the Chinese, the army manoeuvred to eliminate his authority over Chinese affairs. Hardliners carried out their own negotiations with the Chinese, but their object was the creation of that puppet government. By the end of the summer, they had apparently succeeded in luring Wang Ching-wei, a prominent Nationalist figure, into agreeing to lead such a regime. By that time, too, they had persuaded Konoe to establish an Asian Development Board (*Kōain*) to handle all relations – including foreign relations – with China. This step neatly cut the Foreign Ministry and Ugaki out of any peace-making role. Feeling betrayed by Konoe, Ugaki resigned at once, in protest.

Ugaki's efforts may not have mattered much, anyway. Itagaki's position in Konoe's cabinet as Army Minister ensured that the army would further escalate the fighting even as Ugaki had been talking peace. By the end of 1938, the army moved to occupy nearly half of China, a stupendous undertaking under any circumstances. Konoe recognized that negotiations with Chiang were not the answer, so he returned to a hardline position in early November. Using the phrase, 'a New Order . . . in East Asia', he invited China and Manchukuo to join Japan in a pan-Asian partnership of cooperation, development, and anti-communism. These were fine-sounding words, but Konoe's 'New Order' had no place for Chiang Kai-shek or a sovereign China.

Konoe's 'New Order' did have a place for Nazi Germany, though. His stress on anti-communism was calculated to appeal to Adolf Hitler, Germany's leader. Konoe and his new Foreign Minister, Arita Hachirō, were eager to transform the vague 1936 Anti-Comintern Pact into a full alliance with the Nazis directed against the Soviet Union.

There was good reason for an anti-Soviet alliance by late 1938. A border dispute at Changkufeng, where the Soviet Union met Korea and Manchuria, saw fierce fighting between Soviet and Japanese forces in July. To the Imperial Army's chagrin, the Soviets inflicted a sharp defeat on Japan's forces, compelling an unsatisfactory settlement. Itagaki, therefore, strongly supported Konoe's bid for a German alliance.

But how comprehensive should that alliance be? The Germans wanted a wide-ranging agreement that would include Britain and France as possible targets, not just the Soviet Union. Itagaki and the army were perfectly willing, but the Imperial Navy's top leadership was not. Navy Minister Yonai Mitsumasa argued that Britain could not be separated from the United States in any military calculation. Indeed, a German–Japanese alliance was likely to produce an Anglo-American one. If Germany then plunged into a war that involved the European democracies, Japan would find itself dragged into an unwanted conflict with Washington that Yonai thought Japan could not possibly win.

Yonai, however, did not speak for all the navy. Everyone there agreed that Britain and America were indivisible. But while Yonai and other moderates used this point to argue for caution, younger officers, disciples of hardliners Katō and Suetsugu, maintained that the prospect of war with the Western navies provided the best possible grounds for insisting upon a colossal shipbuilding programme for the Imperial Navy. Indeed,

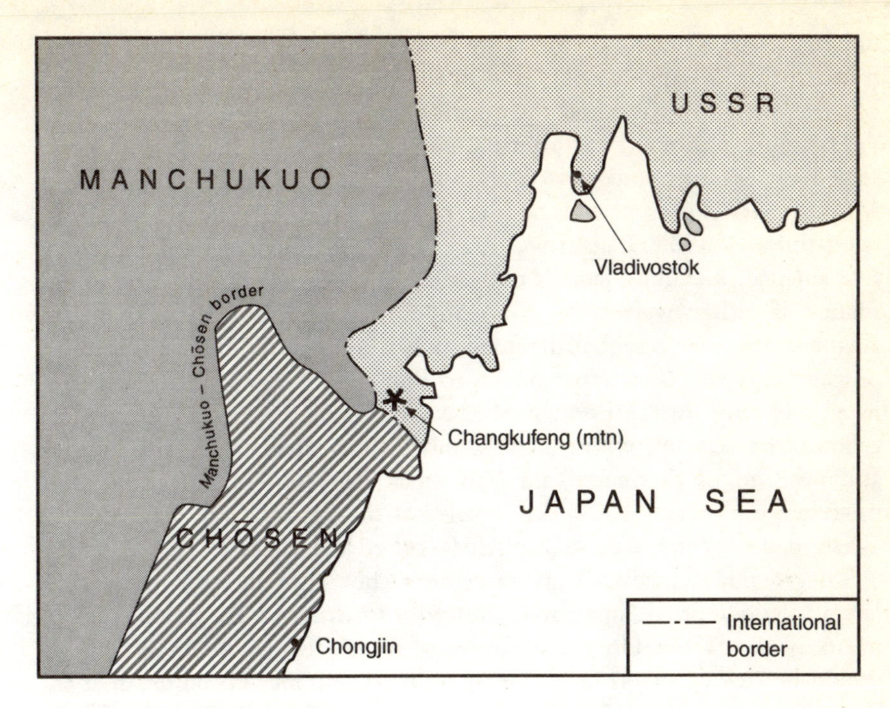

Map 7 Changkufeng 1938

they argued that if the Imperial Navy were unprepared to fight the British and Americans, why did it exist at all? Nor did they shrink from the possibility of armed contest. These men, it should be remembered, had lived most of their professional careers in the shadow of the London treaty controversy. Nearly all of them remembered it as a humiliation for their navy and themselves. They had no recollection, as Yonai and fellow-thinker Yamamoto Isoroku did, of the days of the Anglo-Japanese Alliance. Yonai thus had to fight on two fronts, against both the army and his own subordinates, over the nature of any alliance with Germany.

The showdown came at an inner cabinet meeting on 19 January 1939. By that time Konoe had resigned. He had been unable to bring the China conflict to a favourable conclusion and, while the fighting continued, he was unwilling to go any further with the army's plans for additional domestic reforms. In his place came a veteran reactionary, Hiranuma Kiichirō. The new Prime Minister was conservative in two senses. He did not share Konoe's or the army's hope for an eventual, fundamental restructuring of Japan's polity and economy. But he was virulently anti-communist, so he favoured the alliance with Germany.

Hiranuma's cabinet was a virtual copy of Konoe's. Itagaki and Yonai stayed on as army and navy ministers; Arita continued at the Foreign Ministry. So the arguments at the January meeting differed very little from those of 1938. Itagaki was ready to accept Germany's comprehensive alliance, including Japan's obligation to go to war against Britain and France if either warred on Germany. Yonai personally opposed any alliance, especially one that included such an obligation, but he was under pressure from his own subordinates to find some way to justify increased naval spending. Arita's Foreign Ministry was badly divided between a so-called 'Axis faction', whose members held that Japan shared the outlook and position of Germany and Italy as a 'have-not' nation, and more moderate diplomats, who were appalled at the possibility of war with the Western democracies. Arita himself favoured the alliance.

The meeting produced an agreement, but a predictably weak one. Japan would approve an alliance, but with formal war obligations including only the Soviet Union. If Germany fought Britain or France, Japan would be free to determine its level of involvement. To ensure that the West indirectly understood this limitation, the alliance would be announced as a simple extension of the 1936 Anti-Comintern Pact. In addition, to propitiate the younger naval officers, the inner cabinet agreed to the occupation of China's Hainan island as a possible threat to France's Indo-Chinese colony and Britain's Malayan possessions.

The agreement satisfied no one, including Germany. In early May, Germany proposed that Japan at least agree to assume belligerent status against any countries that might fall under the Comintern's influence – that is, any that might ally with the Soviet Union. Berlin did not insist that Japan commit belligerent acts, but, as critics in Tokyo pointed out, its proposal would have required that Japan go to war whether its forces fired shots or not. Despite Hiranuma's favourable inclination, the navy and Foreign Ministry rejected these terms.

Nevertheless, the world knew of these conversations and many Western capitals feared the worst. Playing on that fear, the Imperial Army used the summer of 1939 to weaken the West's ties to China, hoping to drive Chiang to despair and thus surrender. An incident at Great Britain's concession in Tientsin obligingly gave the army a perfect opportunity to pursue this strategy.

Japan's occupation of China was harsh. The Imperial Army commandeered property and confiscated goods at will. The fighting itself was particularly destructive, with predictable results for supplies of food, clothing, and shelter for Chinese civilians.[6] Some Chinese chose to avoid suffering by collaborating with their occupiers. In April, one collaborator was murdered in Tientsin. The four men accused by the Japanese had taken refuge inside the British concession, a section of the city where Britain retained control under the old institution of extraterritoriality. The Japanese demanded custody of the men. Britain refused to turn them over on the grounds that Japan had presented no evidence against them. In addition, London well understood that to submit to Japanese demands would damage relations with China and harm Chinese morale. This point became perfectly clear after the Imperial Army blockaded the concession and, in June, announced that the blockade would not be lifted even if the four were given up. Britain would have to participate in Japan's new order in East Asia.

The army's new demand was not idle rhetoric. In blunt terms, it was a matter of money and power: the army wanted Britain to recognize only the currency of the Japanese puppet regime as legal. Of course, this would have been tantamount to British recognition of that regime and an abandonment of Chiang.

Britain was in a difficult position. To submit was not only repugnant, it would also be a terrible blow to China's own currency, which was badly suffering from entering its third year of war. To resist, however, invited full-scale confrontation with Japan just when the Polish crisis made war with Germany seem quite likely.

The British therefore delayed, hoping that Japan would impose no ultimatum. None came, in large part because the United States chose July 1939 to give Japan notice that it was abrogating its treaty of commerce and navigation. It was a very serious step, for it meant that in six months Washington would be completely free to shut off all trade with Japan. This was a potential blockade that made Japan's real one against Tientsin pale in comparison.

The navy's occupation of Hainan island and the army's blockade of the Tientsin concession badly strained Japan's relations with London and Washington. But at least there was no open combat. The same could not be said of the Soviet Union, which fought a small war with Japan that summer.

The conflict began between clients. Soviet-sponsored forces from Outer Mongolia clashed with Manchurian troops under Japan at Nomonhan, an obscure location near the western border of Manchukuo. The Imperial Army predictably moved to chastise the Mongolians, prompting Soviet countermoves. This swift escalation led even Army Minister Itagaki to call for restraint, but local Japanese commanders proceeded with a full-dress assault on Soviet positions in late July. The Soviet response included large tank formations and mauled the Japanese badly. Even so, enraged Japanese field officers were ready to raise the stakes higher still, but shocking news from Europe restrained them.

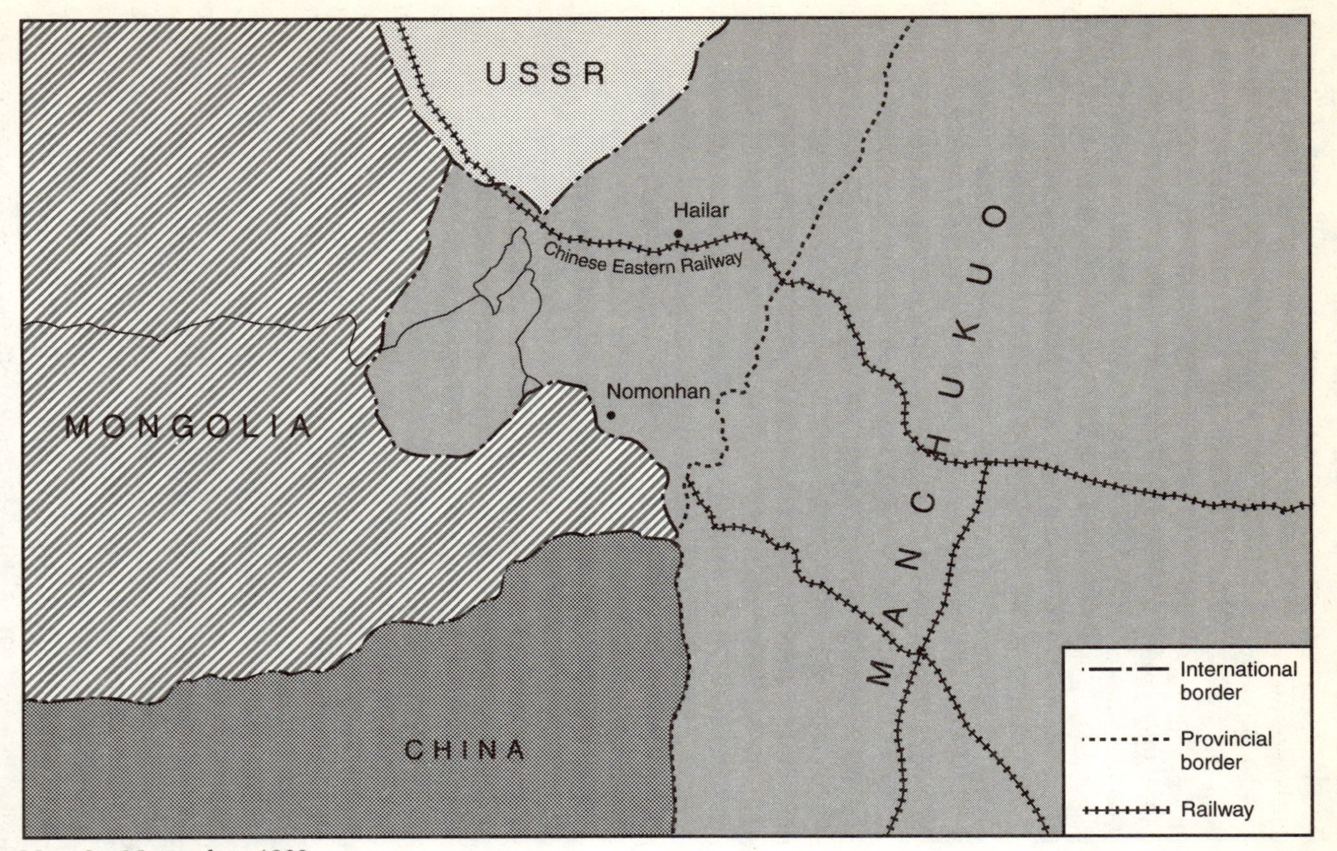

Map 8 Nomonhan 1939

That news was the stunning conclusion of the Nazi–Soviet non-aggression pact in late August and the German attack on Poland, and Franco-British war declarations against Berlin that swiftly followed. Within one week Japan lost a government at home and its way abroad.

Hiranuma and Itagaki had banked heavily on a German alliance to avoid Japan's diplomatic isolation as they and the navy's southward moves proceeded to alienate the other major powers in mid-1939. The Nazi–Soviet pact completely overthrew their calculations and was a complete disaster for their strategy – and Japan's standing – of colossal proportions. There was no alternative strategy to pursue, either. At least for the moment, Japan was very much alone.

Under the new Prime Minister Abe Nobuyuki, a retired general, there was much casting about for a new path. There was some sentiment for following Germany's lead and seeking a parallel understanding with the Soviet Union that, among other things, might lead to an end of Soviet aid to China. But few in the army could tolerate any agreement with their traditional foe and many Japanese conservatives gagged on the idea. Instead, the army set about resolving Japan's problems in its own way. More force would be applied to China in yet another escalation of that draining conflict. At home, Abe would return attention to the army's oft-delayed programme of domestic reforms clearing the path to the new order inside Japan. Foreign Minister Nomura Kichisaburō would buy time for the creation of that order by ameliorating relations with the United States.

There frankly was an element of fantasy to all of the army's initiatives. Its troops had besieged the southern Chinese city of Nanning by early December. The army hoped that Nanning's proximity to French Indo-China might encourage Paris to halt shipments of aid to China through its colony. But France did nothing. China, however, did a great deal, launching counter-offensives in northern, central, and southern China at year's end. These attacks did not regain much territory, but that was not their objective. They succeeded quite well in showing Japan (and the rest of the world) that China still had plenty of fight, and that even Japan's latest escalation remained unable to compel surrender.

In addition, the Nanning campaign made Nomura's efforts to restore cordial relations with the Americans impossible. Nomura hoped to persuade Washington to negotiate a new commercial treaty, which would have barred the United States from applying any economic pressure against Tokyo. But the Americans caustically pointed to the Imperial Army's repeated violations of the Open Door principles in China, and to the army's brutal occupation policies there. As well, the army's ill-timed demand that all belligerent powers' concession garrisons leave China irritated the West, as it applied only to France and Great Britain[7] and thus seemed to be a clumsy attempt to repeat the Tientsin crisis of the summer. The old commercial treaty with America, a sign in its day of Japan's successful emergence as an equal to, and partner with, the West, formally expired in January 1940. Nomura had failed.

Abe failed, too, in his attempts at forwarding the army's domestic reforms. Historians often view Abe as a relative cipher, a caretaker prime minister. This is not how the army conceived of his premiership. Abe himself was unremarkable, perhaps, but he allowed his son-in-law, an ambitious army colonel,[8] to dominate his cabinet. The colonel revived plans for a new, mass political party (along the lines of Germany's National Socialists) to replace the *Seiyūkai* and *Minseitō*. This guaranteed a clash with the old parties and the interests they still represented, such as business and finance. These groups had no faith in the army's policies, political and especially economic. They had good reason to complain. Since the autumn of 1937, the cabinets of Konoe, Hiranuma, and now Abe had gambled that the conflict would be over shortly. To that end, they had diverted scarce resources, from reserves of foreign exchange to fuel supplies, to the military and away from the civilian economy. The predictable result was sharply reduced productivity and increasingly severe inflation. Japanese businesses could not import the materials they needed, such as raw cotton, to remain in production. With fewer shirts or dresses being made, their prices naturally rose. Getting to work became a problem, too, as even public transportation was cut back in the face of fuel shortages. Abe's solution, wage and price controls imposed in mid-October, addressed the symptoms of Japan's economic ills while doing nothing to correct the causes.

The army might have overcome the parties and Japan's business leadership by simple use of force to overthrow the government. But this alternative was never seriously considered, not after the disastrous February 1936 coup plot. Senior army officials understood that they could not run Japan on their own; they had to have the cooperation of the traditional political and economic élites. That was precisely why the army tried so hard to organize a mass party to supplant the old ones – and why it failed. Under Abe, Japan's economy, and its people's lives, worsened appreciably while the Prime Minister could show no gains, in China or elsewhere, to offset these difficulties at home. The fault was not entirely Abe's, nor even the army's. The outbreak of war in Europe had sent prices of war-related materials soaring. These were exactly the materials Japan required to fight China. Even the weather proved uncooperative, with a drought leading to shortages of both rice and hydroelectric power. The army did manage to entice a few politicians into the Abe Cabinet, but these were perennial mavericks such as Nagai Ryūtarō of the *Minseitō*. This was not enough to secure passage of any part of Abe's reform package, and Abe was reluctant personally to challenge the party leaders in the Diet. He suggested calling elections, but the army's leadership surely was correct to realize that this step invited disaster given the level of popular discontent. By mid-January 1940, Abe had to resign.

The army clearly preferred that Konoe return to the premiership and to the struggle for the new order inside Japan and out. But Konoe saw these as linked: until he saw a reasonable prospect of ending the conflict with China, he remained unwilling to attempt a reordering of Japan. The army then turned to another of its generals to replace Abe, but the army's own failures in China and with the Diet made another prime minister from the army impossible. Instead, Admiral Yonai – the scourge of the alliance with Germany – became Prime Minister.

Yonai brought in a cabinet that, under other circumstances, might have reoriented Japan's foreign policy fundamentally. Arita, still Foreign Minister, moved increasingly closer to Yonai's views on the German alliance question. Yonai's chief economic ministers were united in their concern for the army's policy of placing the China campaigns above all other considerations.[9] They were joined by four ministers in the cabinet drawn from the *Seiyūkai* and *Minseitō* in something of a renaissance for these traditional political parties.

This combination was strong enough to block the army's programmes, but the army had sufficient power to stymie Yonai's attempts to change Japan's course. In fact, Yonai's premiership saw increased army activism in both China and Japan. These efforts were interconnected closely. The army could not push its reform programme by itself; it needed to oust Yonai and, for that, it needed Konoe's help. But Konoe would not agree to return as Prime Minister until the China question was resolved. So the army bent every effort toward this end in early 1940.

One such effort aimed to destroy Chiang Kai-shek's legitimacy by creating a rival regime for all China, not just the northern provinces, and luring prestigious Chinese leaders to serve in that regime. One such leader was Wang Ching-wei, one of the *Kuomintang*'s most eminent figures. Wang, distressed by the devastation China had suffered and by the direction Chiang had taken his country, had fled to Indo-China in late 1938 and opened discussions with the Japanese.

Japan did not ignore Wang, but neither did its leaders pursue him seriously – until late 1939. Even then, its terms for Wang were hardly gracious. In exchange for ending the conflict, and allowing Wang to organize a government to replace Chiang's, Wang was to recognize Japanese rights to station troops in north China indefinitely and to exploit the north's resources forever. The Yangtze river valley of central China would likewise be open to Japanese 'economic development'. Finally, in an echo of the most onerous of the Twenty-one Demands of 1915,

Wang's regime was to employ Japanese 'advisers' who would have actual policy-making powers.

These terms gave lie to Japan's hopes for Wang's regime becoming legitimate in the eyes of the Chinese people. Yet the Imperial Army thought them fairly generous, since its forces would evacuate Chinese territory except for the five northern provinces. Wang, having burned his bridges, reluctantly agreed. But two of his senior supporters, critical to his possible success, abandoned him in early January and divulged the Japanese scheme to the world.

It is not certain how seriously the Japanese army itself took these discussions with Wang. Even after he bowed to all its terms, Japan withheld recognizing his regime, preferring to use the threat of recognition to put leverage behind yet another attempt at direct negotiations with Chiang Kai-shek. These went nowhere, until the army finally became disillusioned in late September.

Well before then, the Imperial Army's third initiative toward China – still more force – had combined with Germany's stunning victories in Europe to revolutionize the world and Japan's position in it. Another offensive against China had always been the preferred option for Japan's field commanders, who cared little for diplomatic solutions. In the spring of 1940 their target was Ichang, a strategic city far up the Yangtze river valley, close enough to Chiang's third (and final) capital of Chungking to permit bombing it regularly. Ichang fell on 20 June.

By that time, Hitler had completed his amazing conquests of Denmark, Norway, the Low Countries and, unbelievably, France. To the army, this presented an epochal opportunity to ally with Germany and create a new order for all East Asia, including Japan. Only Yonai and the old party leaders stood in the way. Yonai was ousted on 16 July when the army brought down his cabinet by withdrawing its minister. To eliminate the partymen, the army turned once more to Konoe.

For the first time in eighteen months, Konoe was willing to reassume the premiership. In part, he acted from spite. Konoe had sought the prestigious post of Lord Keeper of the Privy Seal, but had been blocked because officials of the Imperial Court thought Konoe was under the army's influence too much to be in a position so close to the Emperor.

Of more importance was Konoe's growing confidence in the army's Chinese initiatives and his resulting conviction that the conflict with China would soon be over. Once that distraction was removed, Konoe was prepared to lead a new, mass political party, become a prime minister with real power and inaugurate a true new order within Japan to complement the army's new order for the rest of East Asia. Konoe may not have wanted exactly to become Japan's Hitler or Mussolini, but he was impressed with their accomplishments and prerogatives as dictators.

So was the army. To prepare for Konoe's return, its top leaders drafted the 'Outline for Dealing with the Changes in the World Situation' (*Sekai jōsei no suii ni tomonau jikyoku shori yōkō*) in early July. A revealing document, the outline charted an ambitious set of foreign and domestic policies for the new Konoe–army cabinet. Abroad, the chief question was how to best utilize Germany's European victories for Asian advantage. The obvious first step was a formal alliance with Berlin. An equally obvious second step was to pressure France and Britain to prevent any aid from reaching China via their colonies of Indo-China and Burma. Indeed, the army was prepared to occupy some of these British and French possessions – once Germany had commenced the invasion of Great Britain, probably at the end of August. As a bonus, Japan would open negotiations with the Dutch East Indies to wring oil from that colony. Chiang, cut off from all possibility of help from a beaten West, his own army reeling from Japan's Ichang campaign, and his legitimacy shaken by Japan's creation of the Wang regime, would have little choice but to ask for terms.

The 'Outline' had a larger, longer-range goal as well, nothing less than breaking Japan's critical economic dependence on imports from the United States. At first glance, this simply restated the objective of officers such as Ugaki, Nagata, and Ishiwara, who had aimed for decades at a truly self-sufficient Japan. Actually, it was a perversion of their programmes. Through the mid-1930s, these officers had stressed the importance of peace with the West as a prerequisite of Japan's quest for autarky. Only after self-sufficiency was achieved could Japan afford to alienate the democracies. In its 1940 version, however, with Nagata dead and Ugaki and Ishiwara departed and despised, the army favoured conquest of the West's colonies and alienation of the West immediately – before autarky was within sight – in order to achieve that autarky eventually. Presumably, Germany would keep the West humbled and at bay long enough for Japan to complete its submission of China and to advance into the South Pacific unmolested.

The 'Outline' had an imposing list of domestic objectives, too. The cabinet system was to be streamlined; some ministries were to be merged or abolished. The Prime Minister would acquire sole power over cabinet appointments and would be assisted by his own Control Ministry. That new ministry's first business would include a dramatic expansion of governmental controls over the economy.

This was an impressive number of goals, but they were not unrealistic so long as everything went right for Japan. Nothing went right. Konoe and the army tried their hand at new peace talks with Chiang, confident that the Chinese leader would understand that he and the West had lost. Chiang rebuffed them. The prospect of more years of war in China convinced Konoe that he could not push for the domestic reforms in Japan, including a new, mass political party. In September, Konoe did agree to form the Imperial Rule Assistance Association (*Taisei yokusankai*), which technically abolished the old *Seiyūkai* and *Minseitō* parties. But the Association was a fiction existing on paper only. Under the surface, the traditional partymen still held power in the Diet.

The extent of that power became clear in the struggle over Konoe's New Economic Order (*Keizai shintaisei*) in the autumn of 1940. The Cabinet Planning Board, heavily staffed with army officers, proposed centralized governmental control over government-created compulsory cartels for each industry. The business leaders and their allies in the former parties fought the proposal bitterly and successfully. Konoe declined to even bring the proposal before the Diet.

The key, as both Konoe and the army realized, was to end China's resistance. Direct talks had proven fruitless. Perhaps an alliance with Germany and occupation of the northern half of French Indo-China would lead to progress against the stubborn Chiang. Here, the chief obstacle was the Imperial Navy.

Since Yonai's removal from the premiership, naval hardliners had gained influence over Konoe's Navy Minister, Yoshida Zengo. But even they were not about to allow the army to conclude the German alliance without exacting a price of their own. The navy's younger officers were angry that the fighting in China had ensured that the lion's share of Japan's scant resources were going to the army. So they insisted that the German alliance be complemented by a non-aggression agreement with the Soviet Union. This was a rather bizarre twist to the original alliance proposal that had been intended as an anti-Soviet instrument. But it made perfect sense for a navy that wanted more attention and resources paid to its 'Southward Advance' and less to the fighting in China and a possible (army-led) war against the Soviet Union.

At the same time, the navy did not want its Southward Advance to alienate the Americans prematurely, for two reasons. One was simply practical: the navy overwhelmingly relied upon American fuel for its fleet. The second was a matter of faith: the navy was certain that the United States would never abandon Great Britain. To embark upon a fast and furious strike to the south in order to intimidate China, as the army wanted, would entail occupying British territory. The navy thought this step far too risky, or at least too early, since it would mean war with America, and so they blocked it.

The resulting army–navy compromise pleased no one and accomplished nothing of value. Japan would move south, but only in small steps and at a deliberate pace. In early July, the two services agreed to the occupation of only the northern half of French Indo-China. Even this was a minor fiasco. Foreign Minister Matsuoka Yōsuke reached a general agreement with the French Ambassador, but the document failed to indicate specifics, such as how many Japanese troops would enter Indo-Chinese territory. The army, eager to press China, shoved its way into the colony rudely and provocatively. Indeed the Americans were provoked, shutting off exports of top-grade aviation fuel and scrap iron to Japan by the end of September. This American step stiffened the Dutch, whose negotiators in the Dutch East Indies refused nearly every Japanese demand for oil. The stand of the United States and Holland, in turn, confirmed Chiang Kai-shek's own continued resistance to Tokyo.

Alliance talks with Nazi Germany went slowly and uncertainly, too. Hitler preferred that the British sue for peace after France fell in June. An alliance with Japan signed too quickly might have triggered American help for London that would have given the British reason to fight on. Serious negotiations for an alliance did not commence until early September. But by that time, the Royal Air Force's victory in the 'Battle of Britain' had boosted British morale and convinced the United States to transfer fifty warships, designed to hunt submarines, to the Royal Navy. The Tripartite Pact of Germany, Japan and Italy, therefore, intimidated neither London nor Washington. Quite the contrary: Washington's immediate reaction to the new Axis alliance was an embargo of additional scrap metals to Japan, a real blow to Japanese steel production in the years to come.

Japan's leaders had not discounted this result. In their final debates over the German alliance, they had been remarkably candid about the options facing them and their country. In a mid-September meeting with Konoe and the military, Matsuoka argued that Japan finally had to choose between the Axis and the Anglo-Americans. To reconcile with the democracies would return Japan figuratively to 1913, surrendering all the imperial gains in China and the Pacific since that time. In fact, Matsuoka continued, Japan would be in a poorer position, since China would be much more hostile and unified than it had been before the First World War. Matsuoka might have better chosen 1853, given his predictions that reconciliation with the West really meant becoming a sort of Western colony for at least fifty years.

Proceeding with the German alliance, on the other hand, opened up an entirely different future. A tripartite pact with Berlin and Rome might inhibit American aid to Britain in Europe. The British, already under severe German aerial attack, might sue for peace to save their empire in Africa and South Asia. Germany and the Soviet Union would divide up continental Europe, leaving the Americas to the United States – and East Asia to Japan's new order. Germany, and a German alliance, would also prove useful in resolving outstanding Japanese–Soviet difficulties.

It was an exceptionally alluring vision, one Matsuoka would follow over his next year as Foreign Minister. But many other Japanese believed it was a fantasy. Navy Minister Yoshida Zengo and Commander of the Combined Fleet, Yamamoto Isoroku, believed that Britain would continue the war against Germany with increased American assistance. For Japan to ally with Berlin made a war with the United States much more likely and neither Yoshida nor Yamamoto believed Japan could win against the Americans. Pressure from both the army and the navy's younger hardliners drove Yoshida from power (and into a hospital) in early September, making Oikawa Koshirō Naval Minister. But even under Oikawa, the navy valued the alliance less than Matsuoka or the army. It wanted no automatic obligation to come to Germany's defence if the United States became involved in the European war. It wanted more of Japan's resources spent on building up its fleet. The army agreed to both stipulations. So the navy consented to the German alliance by the end of September.

By the autumn of 1940, Japan had made a crucial decision in an unhappy situation. At that critical moment, China had not collapsed, so plans for thorough reforms inside Japan were, at best, delayed. France had been humbled, but at the price of driving America, Britain, China, and the Dutch closer together. The alliance with Germany was at last a reality. But its signing only hardened the 'ABCD' coalition and reduced the options open to Tokyo. In essence, Japan had tied its future to Germany's. If Hitler could induce Britain to sue for peace, if he could reach a permanent European accommodation with the Soviet Union, Matsuoka's grand vision would come true. Unfortunately for Japan, these decisions would be dictated by Berlin's will and power, not Tokyo's. It is remarkable that Japan's September deliberations on the German alliance made much of the loss of sovereignty Japan would suffer if it returned to the Anglo-American orbit, while ignoring the quite significant loss of independence it surrendered through its entry into the German one. But if Japan was determined to achieve security through empire, alliance with Germany was the only realistic option in late 1940.

Even so, the German alliance did not give Japan any initial advantage in its most pressing objective: forcing China to surrender. In fact, the British, emboldened by America's new embargo against Japan, reopened the Burma Road (from British Burma) into China in October, heartening Chiang by restoring substantial Western aid to him. Chiang was also pleased by the complete lack of support that Japan's puppet regime, under Wang, had among the Chinese. He was confident that 1941 would be a good year for China.

The Japanese Government was running out of ways to induce Chiang's surrender, so it turned to Matsuoka's long shot in early 1941. The

Foreign Minister hoped to bring the Soviet Union into a general understanding with Japan and, by implication, Germany. Once Moscow was a permanent partner of Germany, Japan and Italy, Britain, the United States and China would have little choice but to come to terms. He left for a whirlwind tour of Moscow, Berlin, and Rome in March.

Matsuoka's logic was impeccable, but his grip on reality loose. The Germans bluntly informed him that they were gathering troops along the Soviet border. This was hardly a sign of an emerging German–Soviet–Japanese coalition, but Matsuoka blithely ignored the implications. The Soviets gave him a cool reception but, concerned about the situation in Europe, signed a five-year neutrality pact that simply bound each nation to neutrality toward the other if either were attacked by a third power. Stalin showed his lack of trust in Japan, however, by refusing to withdraw any troops from along the Manchurian border. Matsuoka blithely returned to Tokyo in late April, with the German attack on the Soviet Union only two months away.

And negotiations with the United States were already one month old and at a critical juncture due to a well-meaning blunder. Nomura Kichisaburō had been installed as ambassador to the United States at the start of 1941. An old admiral, he had long-standing American connections and hopes that his country's drift toward confrontation with Washington could be reversed. Shortly after his arrival there, two Catholic priests, who themselves hoped to avert war, drew up a 'Draft Understanding' that might be used to resolve Japanese–American differences. American Secretary of State, Cordell Hull, agreed to use the draft as an informal basis for discussions with Nomura, but the Japanese Ambassador's report of the draft to Tokyo left the impression that Hull took the draft far more seriously. Indeed, Nomura's ambiguous reporting led many in the Japanese government to assume the draft was an official American proposal.

As such, it was unbelievably favourable to Japan, arousing both excitement and suspicion within the government. The draft committed the United States to restoring normal trade with Japan and barred it from further embargoes in the future. Japan's freedom to act within its German alliance was to be left untouched. Best of all, Washington would promote a settlement of the China conflict based upon an amalgamation of the Chiang and Wang regimes, in other words, on Japan's terms.

Some Japanese diplomats hailed the Draft Understanding as a guarantee of peace in the Pacific based on a fundamental Japanese–American concord. But the Imperial Army had doubts. To its leaders, the draft was a clever American device to delay a showdown with their country while the United States assisted Britain in Europe. Still, it was worth exploring, as it offered the great prize of ending the morass in China.

The navy was even more chary. It agreed that the Americans wanted to defer any confrontation with Japan. But the navy was more concerned with the United States' colossal naval construction programme, which had begun in the summer of 1940 after the fall of France. The American building was to be completed in large part by the end of 1942. Was it not likely that Hull simply sought to preoccupy Japan with crafty diplomacy while the American navy became large enough to seek Pacific hegemony? Even so, the navy was willing to explore American intentions further, especially since a rupture – leading to an American embargo on oil shipments to Japan – was to be avoided as long as possible.

Only Matsuoka opposed proceeding with negotiations on the basis of the Draft Understanding. To do so, he argued quite rightly, would undermine relations with Germany and, at best, only lead Japan back to a junior partnership with the West. It would be better to cement Japan's German connection by considering an attack on the British at Singapore. With Britain driven from the war, the new order of Germany, the Soviet Union, and Japan could be inaugurated at last. The army, navy, and Konoe refused to consider such an attack, and began to wonder whether Matsuoka was fit to remain as Foreign Minister. Increasing rumours of a German attack on the Soviet Union only reinforced their doubts about Matsuoka's judgement.

But by the end of May they had to call their own assessments of the United States into question. Matsuoka had used his office to delay a response to the 'American' draft understanding for several weeks. Hull, somewhat surprised by the edge in that response, presented what was the first actual American position on any comprehensive settlement with Japan. That position was based on four principles, labelled 'Hull's Principles' but, in fact, the American proposal was founded upon ideas well established in American diplomacy long before Hull had taken office: respect for the territorial integrity of all nations, avoiding interference in other nation's internal affairs, equal commercial opportunity for traders of all nationalities, and refraining from the use of force to overturn the existing order. Applied to Japan in early 1941, these principles dictated a wholesale evacuation of French Indo-Chinese and Chinese territory, perhaps even Manchuria.

The United States also moved to demonstrate its determination. Roosevelt had persuaded Congress to pass 'Lend-Lease' legislation, authorizing very substantial material aid to Great Britain. And he sent a trusted adviser, Owen Lattimore, to Chiang Kai-shek to act as a direct link between himself and the Chinese leader.

By early June, therefore, Japan was increasingly certain that the United States was preparing for a two-front war. 'Hull's Principles' were impossible as a basis of any deal with the Americans. So Tokyo decided to demonstrate its determination by occupying the southern half of French Indo-China as possible preparation for an attack on British, Dutch, and even American possessions in the Southwest Pacific. To be sure, army–navy disagreements over how far and how fast to proceed with this 'Southward Advance' were substantial and even acrimonious. But this time, unlike a year before, the Imperial Navy was in the lead, the army more cautious.

There were two important reasons for the navy's new boldness. The United States' absolutely colossal naval construction programme worried the navy. Japan had an edge in key categories of warships in 1941, but by late 1942 the Americans would have caught up and surpassed it. This, coupled with America's forceful stand of May, raised the alarming possibility that Washington would use its new fleets to crush the Imperial Navy and dictate terms. If there were to be a showdown with the Americans, then, Japan ought to move within the next twelve months, the sooner the better.

A second reason for the navy's haste arose from Germany's attack on the Soviet Union on 22 June. Matsuoka, his grand plan for a German–Soviet–Japanese understanding in ruins, smoothly shifted to full commitment to Germany alone. He favoured an immediate attack northward, against the Soviet Union. There was much sentiment for doing just that within the Imperial Army. But a northward advance meant an end, or at least a delay, in the navy's Southward Advance. It would also mean a new war, one in which the bulk of Japan's increasingly scarce materials would go toward producing the army's artillery and tanks, not the navy's battleships. Therefore, the navy strongly opposed an attack against the Soviet Union and struggled to keep the Southward Advance as top priority for Japan's foreign relations.

It did, but only barely, in the decision of the Imperial Conference of 2 July. With the Emperor's (silent) presence giving the decision the most solemn authority, the cabinet, army, and navy agreed to continue the Southward Advance by occupying the southern half of French Indo-China. Yet the army was free to continue preparations for a possible strike against the Soviet Union by reinforcing its troops in Manchuria.

This compromise satisfied the navy and army, at least temporarily, but Konoe and Matsuoka were uneasy about it. Matsuoka railed that it would backfire, alienating both the Soviet Union and the United States. It would be better to destroy the Soviet menace and then deal with the West. Konoe likewise feared a breach with Washington. Unlike Matsuoka, however, he felt that the Nazi–Soviet war showed the bankruptcy of the German alliance. It was time to explore the possibility of an accommodation with the Americans by reinvigorating the stalled Nomura–Hull negotiations. Matsuoka was aghast at this suggestion, which indeed would have rendered ties with Berlin meaningless and which also, he ominously warned, would mean extreme concessions to China.

As Matsuoka increased his attacks on Konoe's ideas, the army and navy were compelled to take sides. For the navy, Konoe's proposals for negotiations with Washington were difficult. They would threaten the Southward Advance and the navy's construction programmes. But the navy continued to have second thoughts about outright war against the Americans, and so was willing to allow Konoe to see what terms the United States might suggest. In the meantime, the Southward Advance could continue. Matsuoka's proposal to attack north at once and break off talks with Hull, and cancelling the Southward Advance while provoking premature conflict with the United States doubly offended the navy.

The Imperial Army was more ambivalent. Konoe's willingness to make concessions over China struck a raw nerve: the army's hugely costly failure to find a military solution there. And the Americans might even wish to reopen the Manchurian question, intolerable from the army's perspective. However, continuing negotiations would keep the Americans in play – and out of war against Japan – and would serve as a useful substitute for the navy's full-blown version of the Southward Advance. The army could then receive priority for its preparations against the Soviet Union and strike when the time was right. Thus, the army felt that Matsuoka's proposal for an immediate attack into Siberia was premature and his idea of cancelling talks with Washington much too rash.

Matsuoka, therefore, had to go, leaving office on 16 July. But Konoe's strategy of serious negotiation with Washington went nowhere. The army and navy insisted upon adhering to the decisions of 2 July, including preparations for the Southward Advance and war against the Soviet Union. Nor would Konoe be allowed to bargain away Japan's alliance with Germany.

Japan, in short, was preparing for war against Britain, the Netherlands, the Soviet Union, and the United States, possibly at the same time. And Konoe could negotiate with the Americans, but had no power to offer any concessions to them or the Chinese. The complete unreality of this position was brought home by the American reaction to Japan's occupation of southern French Indo-China. Roosevelt called out the Philippines' army and authorized the dispatch of American heavy bombers there. He agreed to the stationing of an air force of American volunteer pilots in the service of China, General Claire Chennault's 'Flying Tigers'. Far worse, the United States froze all Japanese assets in America, effectively prohibiting the purchase of anything American, including oil.

This lightning bolt struck hardest at the Imperial Army, which had not reckoned on such a severe American reaction. An immediate strike against the Soviets was now out of the question: Japan had to have oil, and oil was available only toward the south, in the Dutch East Indies and British Borneo. So the army joined the navy in advocating the Southward Advance. But there were still some differences between the two services. The most fundamental difference was whether the American Philippines should be attacked.

This was not just a fundamental difference, it was critical. The army never meant to abandon operations against the Soviet Union, merely to delay them until southern oil could be obtained. As a result, the army proposed to keep the great bulk of its forces (at least those not still tied down in China) in the north. Its Southward Advance would be a shoestring assault using very few troops against the ill-defended colonies of Britain and the Netherlands.

The navy, on the other hand, had not abandoned its belief that America would never abandon Britain. A Southward Advance that left the Philippines untouched would be an exercise in futility. If war came, the Americans would have to be included at the outset, forcing the army to commit many more troops than its leaders wished. Of course, the war itself would be a far greater war, with the United States in it.

For exactly this reason, the navy's leaders were willing to allow Prime Minister Konoe, and his new Foreign Minister (and former admiral) Toyoda Teijirō, to resume negotiations with Washington. Konoe wanted to do far more than resume those negotiations; by early August he wanted to lead them personally in a summit conference face-to-face with Roosevelt.

Konoe's stratagem was perfectly clear to Army Minister Tōjō Hideki. The Prime Minister hoped to reach an arrangement with Roosevelt that he could present to the army as a *fait accompli*, regardless of its terms. Tōjō and the army's top leadership preferred to veto the summit outright, but were convinced that this would bring about Konoe's resignation. Since previous army attempts to occupy the prime ministership, under Hayashi and then Abe, had worked poorly and promoted domestic disunity, Tōjō and his lieutenants consented to the summit, but only if Konoe agreed not to compromise on certain key issues. Specifically, Japan was to remain faithful to the principle of anti-communism, including the German alliance; Japanese troops would have to be retained in China, at least through the northern provinces, for an indefinite period; China would be governed by an amalgamation of the Chiang and Wang regimes; and Manchuria was to remain 'independent', that is, firmly under Japan's control.

Konoe and Toyoda argued that these terms would make any progress at the summit impossible. In fact, in light of America's insistence that an agreement in principle be reached first, they would block the summit itself. In response, the army was willing to agree to a loosening of Japan's alliance obligations to Germany: Japan would agree not to automatically fight the United States if America entered the European war. But it would not yield on anything else. Furthermore, it insisted that if Konoe could not negotiate a settlement with Washington, he would promise not to resign but rather to lead a united nation into war against the West.

Such a stand ensured that there would be such a war. The Americans were not eager to fight in the Pacific. In their eyes, Germany was a far greater threat to the West's safety than Japan. But the West stood for each nation's right to determine its own destiny and against aggression. Washington could hardly claim to be pursuing these goals if it allowed Japan to colonize and occupy China. So, the army's refusal to yield on China[10] became the critical sticking point with America.

In frank recognition of that fact, the army and navy settled on preparations for war, a settlement ratified at the Imperial Conference of 6 September. The army agreed to a strike against the Philippines at the same time as a drive against British Malaya. And it consented to the navy's request that the diplomats have until 10 October to determine if any settlement with Washington was possible.

Konoe tried with increasing desperation to extricate himself and his country from the descent toward a Pacific war. The very night of the Imperial Conference he had dinner with American Ambassador Joseph Grew, where he pleaded for a summit with Roosevelt. Grew seconded that plea, but neither he nor Konoe had any response to Washington's implicit question: could Konoe convince the Imperial Army to agree to terms the United States would find acceptable? Konoe toyed with the idea of employing the Emperor to enforce the army's compliance, but realized that to use this tactic and fail would shatter Japan's polity, and the sacred institution of the Emperor, beyond redemption. So he simply procrastinated.

By mid-October, Tōjō and the army had run out of patience. Every day that passed saw Japan's oil reserves dwindle and favourable daylight and weather conditions for a successful attack to the south fade. Tōjō ended the navy's vacillation over continuing negotiations by threatening to halt all army preparations for war, a step that would have meant surrender to America's terms. With this, Konoe's last hope for finding an ally to break the army's veto of meaningful negotiations with (and concessions to) Washington failed. He resigned on the 16th.

The army immediately put forward Prince Higashikuni to replace Konoe. Higashikuni had long served in the army, and was a royal prince, that is, with blood ties to the Emperor. He, therefore, was an ideal candidate from the army's perspective, one that would ensure national unity in the struggle to come, and one that would be a perfect figurehead for a renewed drive for the creation of a new order within Japan.

But precisely because Higashikuni was royal, the Emperor's advisers, led by Lord Keeper of the Privy Seal, Kido Kōichi, blocked his appointment. Kido, and the Emperor himself, wanted to be certain that all alternatives to war had been exhausted before the final decision was made. So they needed someone capable of exploring the possibility of resurrecting negotiations with the United States, while keeping the military under control, that is, from vetoing any bargain made with Washington. They had no hope that any civilian could manage the military, so the choice for Konoe's successor quickly narrowed to the serving army or navy minister. Actually, that meant Tōjō, because the army could discipline a recalcitrant navy far more easily than vice versa.

Tōjō became Prime Minister, but with a specific charge from the Emperor. He was not to be bound by any prior policy (meaning the virtual decision for war made on 6 September). Instead, he was to reassess Japan's fundamental position and alternatives from a 'white paper', or clean slate. Tōjō took this task seriously, drawing up a host of questions for his new cabinet to answer. Some assessed Japan's prospects if war came. What was the status of Anglo-American forces in the Pacific? How was the war in Europe likely to evolve in the coming year? Did Japan have enough oil, steel, and other critical materials to have a reasonable chance of prevailing? Was war against only the Netherlands and Britain a possibility? But other questions considered prospects for peace, with Tōjō calling for a consensus on the most generous terms that Tokyo could offer Washington.

Even so, Tōjō and the army were not prepared to yield on certain critical conditions regarding China. Nor could wind and weather conditions be returned to 'white paper'. The army insisted that negotiations could be tried only until mid-November, with war to come a week later. Only Tōjō's personal intervention bought the diplomats until the end of that month. Even so, there was not much time. Nor, because of the army's stance, was there much flexibility in Japan's terms, at least the important ones. A marathon conference of early November decided upon two plans for the Americans, Proposal A and Proposal B. If neither were accepted, war would come.

Because the United States immediately rejected Proposal A, historians have largely ignored it. But the Imperial Army regarded Proposal A as a series of substantial concessions and, as such, a serious bid for peace. In it, Japan agreed to withdraw all troops from China, except for the northern provinces and Hainan island, within two years. Even in north China, forces would leave – the first time the army had consented to any pull-out – though not for 25 years. Japanese forces in French Indo-China would evacuate immediately, ending any threat to British, Dutch, and American possessions. Japan would not be bound to attack the United States if America entered the war in Europe. Japan would agree to apply the principle of non-discriminatory trade throughout China – it would give up its insistence on priority access to Chinese resources – though it would want that principle applied everywhere else (a reference to trade barriers erected by the British Empire, for example). Finally, Japan would commit itself to Hull's other 'principles' insofar as it could.

The Imperial Army regarded Proposal A as exceptionally magnanimous. But Tōjō's Foreign Minister[11] thought it would be difficult to reach such a comprehensive agreement with the United States with time so short. Instead, he favoured Proposal B, which divorced a settlement of the Chinese question from the immediate causes of the Japanese–American confrontation in the Southwest Pacific. Proposal B, in other words, was a limited *modus vivendi* to avert war and buy time for a wider settlement later. It offered Japan's promise not to advance south from Indo-China (and to possibly evacuate southern Indo-China) in exchange for a lifting of American trade sanctions against Japan, American help in obtaining oil from the Dutch East Indies for Japan, and an American pledge not to interfere in Japan's efforts for peace with China. The diplomats thought that Proposal B would be attractive to the Americans because it would free them from the prospect of a Pacific war and allow them to focus on Europe. This was just the problem, objected the Imperial Army. The United States could resolve the European situation in its favour while keeping Japan on a short leash through its hold over Japan's oil supplies until the American navy was so large Japan would have to capitulate without having fired a shot. To even offer to evacuate southern Indo-China was madness, as it would leave Japan unable to consider the option of war if a comprehensive settlement proved impossible. The army wanted no part of Proposal B. But Tōjō overruled his colleagues once again. Both proposals would go to Washington.

There, both failed for the same reason: China. Hull quickly dismissed Proposal A with its provision for north China under Japan's thumb for a quarter century. By the same token, Proposal B was unacceptable, since (as both Chiang and British Prime Minister Winston Churchill reminded the Americans) it would violate the same principles America claimed to be upholding in its growing involvement in Europe. The American navy argued in favour of Proposal B: the United States was unready for a Pacific war and needed time. But Roosevelt would not abandon China and rejected Proposal B.

A relieved Japanese military now proceeded with its war plans and an Imperial Conference was held on 1 December to ratify the decision to open hostilities. There, a still-reluctant Emperor took the unusual step of privately interviewing naval leaders before the conference to inquire about Japan's chances for victory. The navy informed him that Japan would do very well at first and could continue the fight as long as necessary, even if Germany should cease hostilities against the West. In any event, the fleet was already closing on the American naval base at Pearl Harbor and it was time to put trust in Japan's warriors. To ensure surprise, its diplomats would continue the pretence of negotiations until shortly before the Pearl Harbor attack, when they would deliver Japan's declaration of war.

The attack came off brilliantly, destroying the American battleships within hours, although the even more important aircraft carriers were not in the harbour. But the diplomats bungled the decoding and typing of the declaration, which was not presented to Hull until after the attack had begun. The infuriated Americans vowed revenge, but even before their Pacific Fleet began its struggle against Japan's new order in East Asia, important Japanese leaders had commenced their effort against that new order at home.

Notes

1 'Amau' is an alternative spelling of 'Amō'.

2 A typical representative of this group would be Nakajima Chikuhei. These 'new *zaibatsu*', like many of the reformist bureaucrats, chafed under the old economic structure dominated by the traditional, established *zaibatsu*. They welcomed government control of Japan's economy, since they had strong reason to believe that they would be in close alliance with the government, especially the army.

3 Tokonami Takejirō typified this group.

4 The new Finance Minister was Yūki Toyotarō, with solid connections in Japan's top financial circles.

5 This was Nagai Ryūtarō.

6 Ironically, the fighting also damaged Japanese enterprises in China. Under wartime conditions, Chinese who worked in Japanese-owned mills were stigmatized and the mills boycotted. More directly, the war disrupted supplies, not just workers, for the factories. The Imperial Army was far more interested in plunder and production for its own needs. A great number of Japan's private businesses in China withered away as a result.

7 Germany, of course, had no garrisons in China. Japan had eliminated them during the First World War.

8 The colonel's name was Arisue Seizō.

[9] These were Takeuchi Kakichi, head of the Cabinet Planning Board, Fujihara Ginjirō, Minister of Commerce & Industry, and Chief Cabinet Secretary Ishiwata Sōtarō, a Finance Ministry veteran.

[10] In fact, the army would insist in September that any settlement with the Americans and Chinese permit Japan priority access to resources throughout China, not just the north.

[11] This was Tōgō Shigenori.

7
Triumph of the conservatives

The Japan of the Tōjō Cabinet went to war against the West on 8 December 1941 to create a new order in Japan and for all East Asia. It fought two struggles and lost both, well before Japan's formal surrender on 15 August 1945. By then, Tōjō had been out of power for over a year, a victim of the forces for the old order which sought a Japan dominated by a pacifist, conservative oligarchy dedicated to achieving stability within the country through a close partnership between business, bureaucracy, and conservative political parties, and stability abroad through a close association with the United States and the rest of the West. Only the remaining institutional power of the Imperial Army condemned Japan to the sufferings of 1944–5. But these were a necessary, even vital, sort of purgatory, for they destroyed completely the ability of that army to influence Japanese politics once the war was over. During the American Occupation of 1945–52 and the political consolidation that followed within Japan, on the other hand, the power of Japan's pacifist conservatives would become so well established that it would show no signs of weakening as the next century approaches, despite intermittent challenges from the Japanese Left, which favoured a neutral position during the Cold War, and a different group of Japanese conservatives who, not pacifists, preferred that Japan play a more equal (hence more militarized) role with America in conducting the Cold War. Even with the end of the Cold War, and the weakening of the Liberal Democratic Party that housed these pacifist conservatives, there has been little change in the fundamental order within Japan or the basics of Japan's foreign relations.

For Tōjō, war against the West was only a part of the effort to construct a new order for East Asia. At first, in fact, it appeared to be the easiest part. The attack on Pearl Harbor achieved complete surprise and gratifying results. Germany declared war on the United States within days, guaranteeing American distractions in the Atlantic. The vaunted British fortress-city of Singapore fell with astonishing swiftness in mid-

February 1942. By the end of March, key portions of the Dutch East Indies had been seized. By May, lingering American resistance in the Philippines had ended and British Burma had been secured.

But how should Japan rule the East Asia it had conquered? At the start of Tōjō's premiership in October 1941, the Cabinet Planning Board had established an East Asia office to study this issue. It recommended that Japan place top priority upon acquiring national defence resources from the occupied areas: rice from Indo-China, cotton from the Philippines, tin and rubber from Malaya, and, most importantly, oil from the East Indies. To this extent, a measure of Japanese control, direct or indirect, would be necessary over all of these territories. But areas with their own well-developed sense of nationhood, such as Burma and the Philippines,[1] ought to be awarded their independence fairly swiftly (in fact, the Americans had already pledged independence for the Philippines before the war). Besides, these areas had few resources of use anyway. The Indonesians could receive limited autonomy, so long as their oil remained in Japanese hands. Elsewhere, the native peoples were too backward for anything except direct Japanese administrative control. That control should be exercised by the Asian Development Board (already with control of Japan's relations with China and Manchuria), which the Planning Board recommended be expanded into a full-fledged Greater East Asia Ministry.

The Planning Board's proposal was acceptable to the Imperial Army, so long as its top priority – securing resources – remained unquestioned. To that end, the army insisted that, for the time being, the existing colonial administrative structures be retained. Japanese (usually Japanese military) personnel would simply replace departed or imprisoned Westerners to ensure optimal extraction of those resources.

Foreign Minister Tōgō Shigenori vehemently (and predictably) challenged the Planning Board's proposals. They would strip away the primary remaining duties of his ministry. So, in the simple terms of bureaucratic 'turf-fighting', Tōgō was bound to oppose the Planning Board. But he was also concerned about their long-range implications for Japan's diplomacy, that is, for ending the war. However correct or desirable independence might be for various East Asian nations, whether granted immediately or at a later time, their independence could well complicate future attempts to conclude the war with the West. In 1938, in one of its inaugural steps, the Asian Development Board – itself a mere front of the army – had obstructed chances for a settlement with China by torpedoing Ugaki Kazunari's attempt to negotiate with Chiang Kai-shek, a settlement that would have gone a long distance toward bringing the war to an end, or perhaps have avoided the war's expansion against the West in the first place.

Tōgō had a point. A resolution of the China question would have made Western prosecution of a war against Japan much more difficult. It also would have made Japan's claims to be fighting on behalf of Asia against an imperialist West far more believable. Tōjō and the Imperial Army understood this perfectly well. But they had tied their policy in China too closely to the Nanking regime of Wang Ching-wei to alter it in early 1942. On the contrary, they thought the moment of their triumphs over Britain and America the perfect time to call for the amalgamation of all China under Wang and, consequently, the liquidation of Chiang Kai-shek's government. As if to close the door to Chiang all the tighter, the army began a new campaign against him in southwestern China in May.

Tōgō lost all his fights. Not only did the Tōjō Cabinet veto a new approach to China, that spring it also endorsed the creation of the Greater East Asia Ministry, despite opposition from the Foreign Minister and some privy councillors[2] such as Kido Kōichi. The Foreign Minister resigned in protest.

An East Asian area pointedly not included in initial Japanese plans was Siberia. Japan's policy toward the Soviet Union in early 1942 was partly opportunistic, partly contradictory. The Soviet Union was the traditional enemy of the Imperial Army. No one could deny the very great advantages for Japan's security if Siberia were no longer under Moscow's control. The army had been eager to prepare for an attack north in the summer of 1941, prior to the American oil embargo. It had prepared for a Southward Advance that would be brief and relatively small, so that a strike against the Soviets would be feasible soon after the West had been defeated. But, in the spring of 1942, the Imperial Army had little desire for war with Moscow, unless Stalin's regime was at the point of collapse and victory over it would be swift and certain. The army's reluctance arose from its continuing inability to resolve the China problem. To embark upon another grand continental adventure while so many troops were still bogged down in pacification duties in China seemed unwise. As well, the army was uncertain how to address the navy's remarkable successes throughout the spring. The admirals were pressing for army troops to occupy Ceylon, in the Indian Ocean, or Australia, or for a second strike into the Central Pacific, this time for Midway Island.

But the most fundamental cause for the army's caution was its estimate of the German–Soviet struggle. By early 1942, the Imperial Army was increasingly certain that Germany would not win. The likeliest outcome was a stalemate. But a stalemate could work to Japan's advantage. If Germany and the Soviet Union called a ceasefire, perhaps with Japanese mediation, the West would have little option but to cease hostilities in Europe and, in consequence, in Asia too. It might well fall to the Soviet

Union, in turn, to mediate peace between Japan and the West. Ironically, the soon-to-resign Tōgō shared this view. In fact, he hoped to be active in speeding a German–Soviet truce through a Japanese initiative. Needless to say, any Japanese attack upon the Soviet Union would make a shambles of this strategy. Accordingly, the cabinet refused a German proposal in July to undertake joint operations against the Soviets, despite Germany's extensive efforts to persuade Tokyo that Stalin was on the brink of collapse. Instead, by the autumn, the army would attempt to mediate a Nazi–Soviet ceasefire in blithe ignorance of Hitler's hatred of the Soviets, so deep that he would never consider a truce with them under any circumstances.

By that time Japan's own war situation had taken a dramatic turn for the worse. The Imperial Navy had attempted to pressure Britain and the United States into accepting a settlement confirming Japan's mastery of the western Pacific through the straightforward method of brute force. Every effort failed. A massive raid by Japanese aircraft carriers against British India damaged much shipping but occupied no territory: the Imperial Army, preoccupied with China and the Soviet Union, had refused to provide any accompanying troops. The navy's attempt to seize the southeast tip of New Guinea at Port Moresby, in order to threaten (and perhaps later carry out) an invasion of Australia, was thwarted by the surprisingly swift interposition of American aircraft carriers in the Battle of the Coral Sea. Most famously, and most disastrously, the Japanese navy's huge operation against Midway Island, designed to destroy the remnants of the American Pacific Fleet, instead saw the loss of the best of Japan's aircraft carriers and their crack fliers. It was a blow the Imperial Navy would never recover from.

As if to confirm that judgement, the Americans invaded the Solomon Islands at Guadalcanal in August 1942. The Imperial Army, finally realizing the importance of the fighting in the Pacific, sent substantial forces to eject the Americans. But by the start of 1943 it was clear that Guadalcanal – and Japan's initiative in the Pacific – had been lost forever.

Despite strict wartime censorship, which was entirely in the hands of the military regarding the war's situation, the truth about Japan's plight could not be hidden from all civilians. Kido was the Emperor's liaison to Japan's Supreme Command and, therefore, aware of the awful truth. He was appalled by the prospect of losing the war, which might jeopardize everything that Japan had achieved since the Meiji Restoration. Kido realized that even the institution of the Emperor might be in danger if Japan lost. He helped organize a cabal of leaders, from former Prime Ministers such as Konoe Fumimaro to senior diplomats such as Yoshida Shigeru. All were concerned about the social chaos that would follow total defeat, especially the possibility of a communist Japan. These conservative dissidents tried to find just the right moment to appeal for peace: when Tōjō and the army would have been driven from power through military defeat and disgrace yet before Japan was so beaten that it would succumb to the communists. Ideally, Japan's polity – and its territory – would be restored to their conditions of prior to 1931.

Konoe, Kido, and company knew that their vision of peace depended upon the tolerance, in fact the active cooperation, of the West. Although dismayed by Roosevelt's call for unconditional surrender at the Casablanca Conference at the start of 1943, they were fairly certain that Western assistance was assured. In part, their confidence stemmed from their general belief that the West would not permit a communist Japan. But, more specifically, these Japanese leaders closely followed the Anglo-Americans' treatment of Italy after its surrender, where anti-communist Marshal Pietro Badoglio and King Victor Emmanuel were awarded (at least temporary) control of the government. At the same time, this 'peace group' enjoyed its first concrete success, as Shigemitsu Mamoru, a close friend of Kido's, replaced Tōjō's choice as Foreign Minister in April 1943.

Shifts in Japan's relations with the rest of East Asia swiftly followed. Shigemitsu pushed for an immediate grant of independence to Burma and the Philippines, as well as Malaya and Indonesia. Shigemitsu hoped that these gestures would show Washington that Japan respected the principle of self-determination, a cornerstone of American foreign policy. As well, the army allowed Shigemitsu to commence discussions with Chiang, in which Japan offered to remove its forces from China and renounce its rights to station them there in the future. To be sure, the army did not allow much else. It refused independence for Malaya and Indonesia, and insisted that Chiang end his alliance with the West as a precondition for Japanese withdrawal. Nevertheless, these were important first compromises, fuelled primarily by the army's realization that Germany's defeat was only a matter of time.

Ironically, the 'peace group' moved slowly for the balance of 1943, and into early 1944, precisely because its members thought that their efforts to weaken the army's influence with Japan would be made far easier after Germany surrendered. They also wanted time to proceed toward achieving the independence of most East Asian countries. If the West recognized the new countries once the fighting stopped, Japan might well inherit an Asia more susceptible to its influence after the war than before.

These hopes were largely dashed by the end of 1944. The 'peace group' saw the successful removal of Tōjō from power, but failed to achieve any of its other aims. Perhaps most important in this regard was Shigemitsu's inability to come to terms with Chiang. His attempt had slim chances at the outset, given the army-imposed precondition of an end to China's alliance with the West. But the West itself played a role. At the Cairo Conference at the end of 1943, with American President Franklin Roosevelt and British Prime Minister Winston Churchill, Chiang had demonstrated unshakable fidelity to the alliance, and had received a declaration that Japan was to be stripped of all its imperial territories, in China, Manchuria, even Formosa and Korea. By the summer of 1944, the army again was in control of Japan's Chinese policy and again tried the failed methods of 1937 and 1938: sheer military muscle. At the same time, events in Italy gave pause to the 'peace group', as the Americans (with Britain's grudging consent) not only compelled Badoglio to resign but also dictated Victor Emmanuel's *de facto* abdication as monarch; his son, Umberto, would be installed as 'Lieutenant General of the Realm', with rather uncertain prospects. The Italian monarchy itself was not dead, perhaps, but its future was hardly assured for the postwar period. The implications for Japan's imperial institution were troubling.

At the same time, the 'peace group' did not fare well in Japan. Tōjō resigned as Prime Minister in July 1944 after the fall of Saipan. By that time, the army had completely alienated the old conservative élites and even its erstwhile reliable allies such as Kishi Nobusuke, a reformist bureaucrat from the Ministry of Commerce & Industry who had done much pioneering work in the development of Manchukuo for the army. The *Seiyūkai* and *Minseitō* had been dissolved in 1940, but their leaders and members still held many seats in the Diet, waiting for a return to power once the army was gone. And the Emperor's advisers, such as Kido, were prominent in the 'peace group'. Even so, these groups combined were not able to unseat Tōjō. Only American military force had been sufficient to do that. And even that force was not enough to remove the army entirely from power. The result was thirteen months of horrendous punishment for Japan.

By mid-1944, the Imperial Army was no longer able to achieve its new order for East Asia or its domestic changes in Japan. But it still had its institutional prerogatives, especially its 'right of supreme command' (*tōsuiken*). The army's leaders understood that Japan's surrender would mean its obliteration, so they resolved to continue the war at whatever the cost to the country.

The immediate question was who should follow Tōjō as Prime Minister. Since no cabinet could exist without a minister from the army, the selection was difficult for the 'peace group', which agreed that any new Prime Minister had to have the army's support – meaning he had to be an army man. The choice fell to former general Koiso Kuniaki, who pledged to continue the war. The 'peace group' made only modest

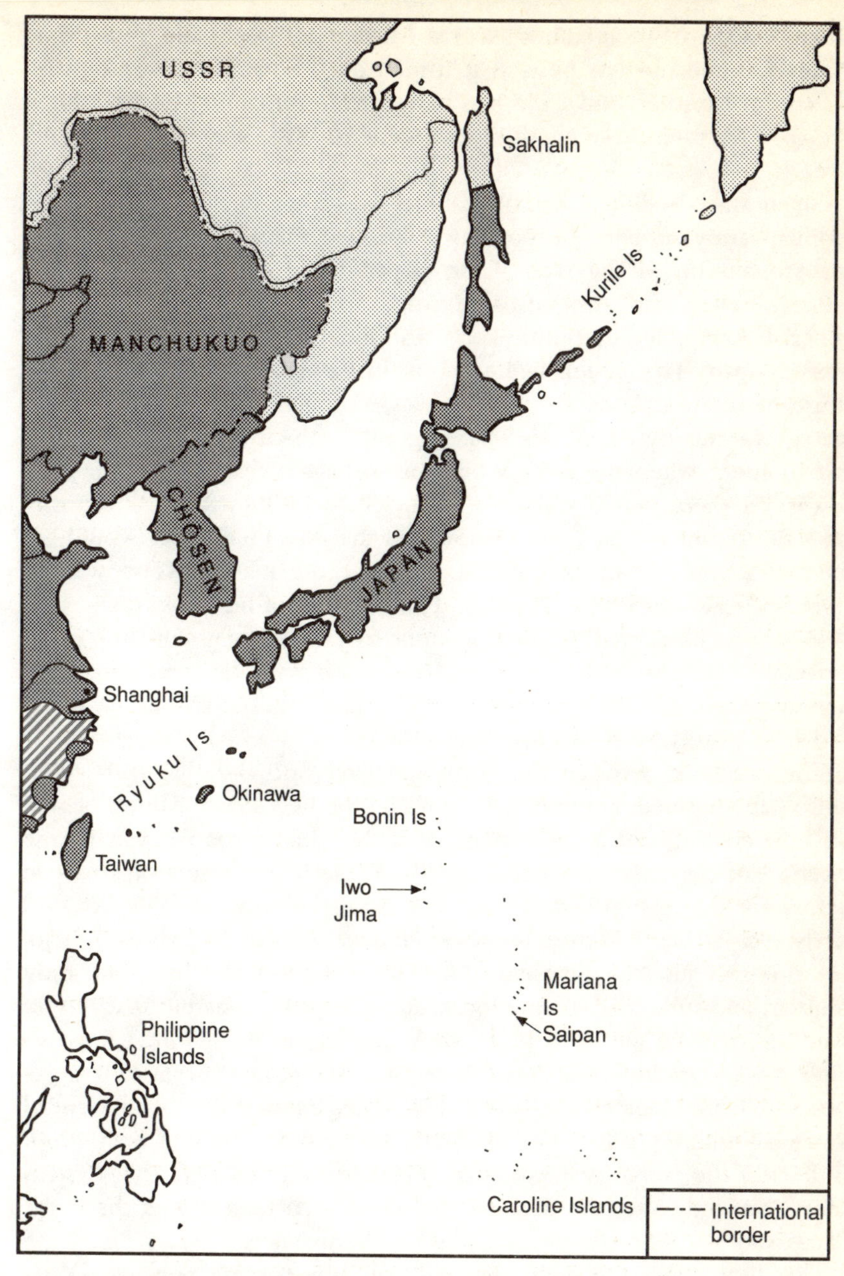

Map 9 Western Pacific 1944–5

progress. One of its members, Yonai Mitsumasa, was made Vice-Prime Minister (a meaningless post, as it turned out), Shigemitsu was kept on as Foreign Minister, and Tōjō was barred from remaining in the cabinet as Army Minister, though that post went to Tōjō's choice, Sugiyama Gen.

Given this deadlock, Koiso's cabinet was a predictable disaster. To maintain army support, he publicly continued to call for Japan's vigorous prosecution of the war. Koiso hoped for a decisive battle in the autumn, when the Americans returned to the Philippines. But the Imperial Army, by that time, was not interested in a decisive battle outside Japan. That action had to occur in the home islands, so that the future of Japan and the army were decided together. It was the perfect strategy to employ against the 'peace group', which was still waiting for that moment when the army would be so beaten that it would lose all power yet there would still be a Japan left to make peace. At virtually the same moment that Koiso announced that the Philippines would see decisive action, the army was abandoning those islands. Koiso was, in short, an irrelevant Prime Minister. The occasion of his fall in early April 1945 was an ill-starred personal attempt to join Chiang Kai-shek in an anti-communist front at the price of abandoning the Japanese-sponsored Nanking regime. This hopeless, almost pathetic initiative only symbolized how powerless Koiso had been.

The deadlock between the army and everyone else that had led to Koiso's appointment continued in April. So the new Prime Minister could not be a civilian, much less a member of the 'peace group'. Even so, the selection of the ancient Admiral Suzuki Kantarō was a sort of symbolic victory for that group. Suzuki was not from the army and, in fact, had barely escaped death during the abortive *coup d'état* of 26 February 1936. But this was just the problem. Tōjō threatened to withhold an army minister, and submitted a name only after it became apparent that Suzuki would form a cabinet without an army minister in it, inviting the possibility of a constitutional crisis. Even so, the army's eventual choice, General Anami Korechiko, was a confirmed bitter-ender. And General Umezu Yoshijirō, a firm ally of Tōjō, was Army Chief of Staff. Both made clear the army's willingness to discuss peace terms with the Western allies – but only after the decisive battle had been fought over the home islands and horrific casualties inflicted on the invaders.

The 'peace group' and its allies did not have enough power on their own to compel the army to give up. Instead, they had to rely upon the Americans for help. They received little. In Washington in the summer of 1945 a different struggle was going on. Senior officials in the State Department with experience in Japan, such as former ambassador Joseph Grew, argued exactly as the 'peace group' hoped they would. Grew maintained that Japan ought to be a Pacific bulwark against communism. The militarists had to be removed, of course, but the moderate civilians (that is, the 'peace group's' members) ought to be put in power and the Emperor system retained as a symbol of legitimacy and (anti-communist) order. Grew drew up the first draft of America's ultimatum to Japan, to be issued at the forthcoming Potsdam summit conference of victorious powers, with precisely these aims in mind. His draft called for the unconditional surrender of the Japanese military, not the government. And it offered assurances that the Emperor system would be kept intact. Unfortunately for the 'peace group', Grew was overruled by hardline Americans who mistrusted the Emperor and, for that matter, all Japanese who had held any power since the Meiji Restoration. The Potsdam Declaration turned out to be a much harsher document, insisting that the only postwar government acceptable to the West was one formed by the 'freely expressed wishes of the Japanese people'.

This stance made it far easier for the Imperial Army to attach its fate to Japan's and argue for fighting on, even after word came of an atomic attack on Hiroshima on 6 August. Cruelly but accurately, the army's representatives, Anami and Umezu, argued that Japan had endured terror from the air for months (including a fire attack on Tokyo in March that took more lives than the atomic bomb did initially). Japan need only hold out for the West's ground attack, inflict huge casualties, and secure acceptable terms. Significantly, the army called for four conditions: (1) Japan would disarm and demobilize itself. (2) Any trials for war crimes would take place under Japanese law. (3) No occupation forces would be stationed in the four home islands. (4) The Emperor system would be preserved.

This was an exceptionally clever amalgam. The first three conditions would all be instrumental in ensuring that the integrity of the Imperial Army would be preserved. The last, the Emperor issue, was a key addition, for the 'peace group' was itself especially insistent on retaining the imperial institution as a means to avoid a communist revolution in Japan and, not coincidentally, to serve as the foundation of a resurgence of conservative civilian authority.[3]

The army had calculated well. Even the intervention of the Soviet Union on 8 August, the swift loss of Manchuria, and the dropping of a second atomic bomb on Nagasaki a day later, failed to shake its insistence on these terms. It appeared either that the army's terms would be met, or Japan itself would be destroyed along with the army. But the army had overlooked one thing. As Kido and other 'peace group' leaders agonized over how to untie the army's umbilical to Japan, Emperor Hirohito determined that the Emperor system itself could be sacrificed to save the country. At two critical cabinet meetings on 10 and 14

August, he declared his satisfaction with the wording of the Potsdam Declaration and insisted on peace at once. Even the army could not overrule an imperial wish, but in case its commanders had such thoughts, Hirohito dispatched three imperial princes to army field commands to ensure surrender. Another prince, Higashikuni Naruhiko, was made Prime Minister to greet the American occupiers. The Imperial Army was undone. Now the conservative 'peace group' would face a second and, as it turned out, far easier struggle to implement their plans for Japan with General Douglas MacArthur, top occupation commander, and his superiors in Washington.

It would seem that by 1945 Japan's foreign policy had come full circle from 1853. The West's military might had returned to Japan's shores, this time led by MacArthur instead of Perry. In 1945, as in 1853, Japan's earlier policies had failed to cope with the Western challenge, and Japan's sovereign existence appeared to be in jeopardy. In many ways, of course, Japan was in a more difficult position, since Western forces clearly meant to occupy the home islands themselves for an indefinite period of time and with an indefinite agenda during their stay.

In other respects, however, Japan's situation was more favourable in 1945 than it had been nearly a century before. The West was not a mystery: Japan had a corps of highly trained diplomats, technicians, and business leaders with excellent knowledge of, and often close personal contacts in, the United States and Europe. As importantly, it was clear that the Imperial Army and Navy would cease to play any role in the postwar era, ensuring that Japan's stance toward the West rested on a broad consensus favouring cooperation.

It remained to be seen on what terms that cooperation would take place. Japan's top priority was to ensure the survival of the Emperor system and avoid the rise of a radical Japanese Left. Closely related to this objective was keeping the substance of the entire Meiji political system (except the military) intact.

Japan commenced its postwar diplomacy to achieve these goals only hours after its formal surrender aboard the USS *Missouri* on 2 September 1945. Foreign Minister Shigemitsu Mamoru protested MacArthur's plan to issue American military currency for use in Japan. According to Shigemitsu, the Potsdam Declaration pledged that the Japanese Government would continue to exist, so Japanese currency – the yen – ought to remain in circulation. MacArthur agreed.

Although Japan would maintain its currency, it was not clear whether it could keep the Emperor. Japan's leaders were all too aware of American public sentiment to try Hirohito as a war criminal (or simply dispense with the trial and hang him straightaway). MacArthur himself tipped them off on Washington's plans to do much more than simply remove the military from the Japanese polity: his office, SCAP (Supreme Commander Allied Powers), was filled with American reformers eager to redistribute land, abolish the powerful industrial–financial corporate groups (the *zaibatsu*), and thoroughly decentralize everything from local government to police and education.

Conservative Japanese leaders feared that these measures would destroy their power and might communize Japan. Shigemitsu's protests led to his removal from office (in fact, he was arrested as a war criminal at the insistence of the Soviet Union), furthering these concerns.

Fortunately for Japan, Shigemitsu's replacement was the clear-headed and crafty Yoshida Shigeru, the dominant figure in Japan for the next ten years. Yoshida realized that his country lacked the power to frustrate American reforms completely. But if it appeared to willingly accept those reforms, it could temper them, perhaps even use them to rebuild a conservative Japan. Under Yoshida, Japan would agree to undertake whatever steps the Americans wished, so long as the undertaking remained in Japanese hands. In this respect, Yoshida skilfully played upon MacArthur's racialism. He self-effacingly argued that the Japanese were not of the West, and so had not yet matured into Western institutions. Japan's errors, Yoshida maintained, were those of an adolescent still learning, quite unlike the deliberate perversion of Western ideals and morals indulged in by the Nazis.

Yoshida was willing to humble himself and his country so long as he could preserve the Emperor system at the core of Japan's polity. He encouraged Hirohito to see MacArthur personally, and asked MacArthur to agree to such a meeting. The result was Hirohito's visit of 27 September and the most famous photograph of the Occupation era: a cocky MacArthur in informal uniform, hands on hips, towering over a reserved Hirohito in Western tuxedo. The symbolism was unmistakable.

It was also objectionable to Japanese conservatives who did not share Yoshida's preference for accommodation. The Home Ministry held an angry press conference, warning the Japanese that anyone who advocated the overthrow of the Emperor system would be arrested. The Americans thereupon fired the Home Minister and ordered the release of all prisoners who had been arrested for such 'political' crimes. Conservatives were appalled, since many of those freed were Japanese communists, and Higashikuni resigned in protest.

Shidehara Kijūrō, the pro-Western diplomat of the 1920s, became Prime Minister. But the real power lay with Yoshida. Yoshida knew that the Emperor's meeting with MacArthur had saved the system. He now moved to confirm that salvation. The Americans insisted that Japan rewrite its Meiji Constitution to end all possibility of a future reversion to militarism. Yoshida was delighted to comply, so long as the new constitution also enshrined the Emperor system and conservative hegemony, that is, so long as the Japanese Government could write the new document.

MacArthur was willing to permit this, but the Japanese chose Konoe to undertake constitutional revision. In Japanese eyes, Konoe was ideal: he was well connected with all the conservative élites and determined to avoid a communist Japan. But his selection was diplomatically unfortunate, as Americans remembered Konoe as the Prime Minister of the prewar government until just weeks before the Pearl Harbor attack. Public pressure forced MacArthur to disavow Konoe's efforts by early November, and the embittered prince killed himself a month later to avoid trial as a war criminal.

By that time the Americans had grown insistent upon a thorough and swift constitutional overhaul. Shidehara stoutly resisted any alteration, holding that any constitutional change imposed upon Japan would be contrary to the Potsdam Declaration's assurance that its postwar government was to arise from the free will of the Japanese people. Yoshida realized that even MacArthur would not be so accommodating. So he encouraged the appointment of a (conservative) Japanese legal scholar, Dr Matsumoto Joji, to replace Konoe.

It was one of Yoshida's few miscalculations in dealing with the Americans. MacArthur's headquarters insisted upon holding elections by March 1946 (to replace the Diet members elected in 1942) and had to have a new constitution in place well before that time. When the Americans learned that Matsumoto's draft constitution continued to accord a measure of sovereignty to the Emperor, they took matters into their own hands and drew up what has ever since been called the 'MacArthur Constitution'.

Shidehara, Yoshida, and other conservatives were dismayed by the new document, with its firm declaration that sovereignty resided with the people and radical provisions such as the rights of women to vote, of labourers to organize unions and, most spectacularly, its renunciation of war and the military in Japan for all time. They urged MacArthur to consider revisions. He rebuffed them, pointing out that the new constitution allowed for the Emperor to continue in a symbolic role and, more to the point, not be tried or hanged as a war criminal.

Yoshida was convinced that the MacArthur Constitution provided adequate safeguards to protect the conservative order. Indeed, he applauded the 'no-war' clause, which not only ensured the permanent removal of the Japanese military élite from politics but also would serve as a guarantee that Japan would have to accept protectorate status from the United States. Why did Yoshida favour a seemingly subordinate role for his country? In brief, he wanted to be certain that his brand of pacifist conservatism would run the country. Yoshida certainly opposed communism. But he also feared other conservatives, such as *Seiyūkai* leader Hatoyama Ichirō, who wanted Japan to pursue a more independent foreign policy in the future. Yoshida believed that a firm, even slavish, alliance with the United States would ensure domestic order and, as importantly, economic assistance and a wealthy market for future foreign trade. His support for the MacArthur Constitution was crucial in ensuring its passage by the newly elected Diet in October 1946.

Yoshida's judgement was quickly put to the test. Organized labour quickly became a political force, supporting the Japanese Socialist Party. On the right, the old *Seiyūkai* and *Minseitō* leaders were 'purged' (barred from public office) by American reformers. Although the Americans barred a general strike that labour threatened for February 1947, the Socialists captured more Diet seats in the April election than any other party.

Yoshida was fairly unconcerned. The purges were mainly directed at the more militant conservatives, such as Hatoyama. In fact, they had been instrumental in seeing Yoshida named head of the *Seiyūkai*, which came in second in the election. Yoshida rightly judged that the Socialists, divided between more centrist union leaders and left-leaning intellectuals, would be unable to form a cabinet. But he was shocked when the Americans, believing that the old *Seiyūkai* still had too many military connections in its past, prohibited his forming a coalition with the Socialists to rule Japan. Instead, the occupation authorities orchestrated an alliance between the Socialists and the old *Minseitō* under Ashida Hitoshi. Yoshida was dumbfounded: the Socialists favoured a foreign policy of non-alignment, not alliance with Washington. And he was appalled as the Americans extended their 'purges' into the business community. As corporate leaders were forcibly retired and industrial combinations dissolved, Yoshida was faced with the prospect of a Japan in the hands of leftists at home and without a protector (or foreign aid or markets) abroad. By the end of 1947, Yoshida and the other pacifist conservatives were facing failure.

They were rescued by the coming of the Cold War to East Asia. Soviet–American relations had been deteriorating since mid-1945. The first crises had occurred in Europe and the Mediterranean, diverting the attention of Washington's top policy-makers from Asian concerns. By the start of 1948, however, they were gazing across the Pacific once more. Chiang Kai-shek was on the verge of losing his struggle against the Chinese communists. As a result, the United States could not hope for a friendly China to contain Soviet ambitions in East Asia. Japan was the logical choice to replace the Chinese, in no small measure because it had demonstrated such great strength during the Second World War. In the spring of 1948, George Kennan, the architect of America's containment policy, visited Japan. He was appalled by what he found. Japan's economy was a shambles and, hence, a potential breeding ground for leftist revolution. Just as badly, the American purges and MacArthur's constitution had rendered the Japanese Government helpless in the face of any rising Red tide. The Home Ministry had been dissolved. There was no national police force, nor, of course, an army. Kennan's recommendations were clear and urgent: Japan needed an immediate economic recovery programme and immediate rearmament (with an immediate alliance to the United States).

No one in Japan better understood the implications of Washington's new priorities than Yoshida, who was still smarting from MacArthur's earlier manoeuvre to deny him the prime ministership. Yoshida realized that he could use America's cold war belligerency to weaken MacArthur's support for the Japanese socialists and their left-wing economic programme. He could use Kennan's desire for a Japanese recovery to end the purges against his allies in business and the bureaucracy and to obtain American economic aid. But Yoshida had no desire to implement the other part of Kennan's programme: Japanese rearmament. In part, Yoshida believed that re-armament would simply stifle any long-term progress toward prosperity for Japan. Yet Yoshida also feared a resurgence of the military as a political force within Japan, especially a military allied with many of the old political party professionals who had collaborated with it during the war and who, not coincidentally, would challenge Yoshida's leadership of the old *Seiyūkai*. Yoshida, in short, was a conservative, but also a pacifist. He was willing to tie Japan to the United States in exchange for protection and access to world markets, but not a Japan that would arm itself for the coming struggle in the Cold War.

Before implementing his programme, Yoshida again had to become Prime Minister. Scandal and intrigue aided him in this regard. As Japan's unions pulled the Socialist Party to the left, Ashida kept his coalition cabinet in power by buying votes in the Diet. Yoshida's allies discreetly informed those Americans in MacArthur's headquarters who agreed with Kennan's new cold war approach to Japan. Eager to oust the socialists, they prodded the Justice Ministry to expose the bribes. Ashida fell by October and Yoshida became premier.

By the end of 1948, Yoshida had forged a marvellous alliance with the American cold warriors, an alliance that reduced MacArthur's personal role in the American Occupation to virtual insignificance. One key new player was Joseph Dodge, an American banker who arrived in Japan as Yoshida was reassuming power. Dodge's chief objective was to build the foundation for a swift Japanese economic recovery, just as Kennan had argued. The resulting 'Dodge Line' was strong medicine. Dodge argued that Japan had to become competitive in international markets to survive, since it had no resources at home. But Japan's economy was weak because it depended too much upon government subsidies (often funded with American aid, to Washington's growing resentment). The best way to induce competitiveness was radically to reduce government handouts. Many companies would go bankrupt, but these would not have survived on their own anyway. Even more directly, many government workers would have to be fired. As well, it was time to stop hobbling Japan's economy by keeping its best business leaders out of power. The 'economic purges' would not just be stopped, but reversed.

These economic effects fitted in perfectly with Yoshida's political plans, as did a rash of American orders recentralizing control of Japan's police forces and moving to restrict the ability of governmental employees to strike. With great pleasure, Yoshida oversaw large reductions in the size of governmental unions (such as the rail union) which often were strongly left-wing. Union-inspired protests, and sometimes even sabotage, were met by police crackdowns and, when necessary, the use of gangsters to intimidate union leaders. Through it all, Yoshida portrayed himself as Japan's saviour from chaos and communism. He also called elections for January 1949, in which his party crushed all others.[4] Many of the new Diet members in Yoshida's party, moreover, were his handpicked supporters.

Having secured his power within Japan, Yoshida now turned to the nettlesome issue of Japan's position in the new world of the Cold War. America's strident anti-Soviet stance had brought him immediate benefits, but also one danger. Even as Yoshida triumphed over communism in Japan, the United States urged him to join it in the struggle against communism abroad, especially after China fell to the communists in late 1949. Yoshida wanted Japan to play no part in this spreading conflict. Japan was to benefit from America's efforts against communism, not participate in them. That meant, above all other things, that Japan was not to rearm.

Under ordinary circumstances, it would have been difficult for Yoshida to resist American pressures for Japanese rearmament. After all, American forces still occupied Japan, and the still-fragile Japanese economy was heavily dependent upon American assistance. But Yoshida had two powerful forces of his own. The 1947 Constitution's 'no-war' clause gave him a perfect excuse to resist American calls for a reborn Japanese military. And it was clear that amending that constitution (which required a two-thirds' vote in both houses of the Diet plus a majority of the Japanese people in a popular referendum) would be all but impossible given the steadfast opposition of the Japanese Socialist Party. It was ironic, but Yoshida and his pacifist conservatives, men who strongly opposed the Socialists' domestic programmes for Japan, were eager and effective allies in blocking any revision of the constitution's 'no-war' clause.

However, if Japan would have no military forces, how was it to be protected? This question became especially pressing in early 1950, when the new (communist) People's Republic of China announced an alliance with the Soviet Union specifically directed against Japan and the United States.

The Japanese socialists had a quick answer to the security question: Japan should not participate in the Cold War at all. Instead, it should declare itself neutral, such as Switzerland or Sweden had done in Europe, turn inward, and concentrate on domestic change.

Yoshida and his pacifist conservatives thought this approach entirely unrealistic. To Yoshida, a man schooled in the history of Japan's foreign relations, Japan could not shut its doors to the world in 1950 any more than it could have in 1853. The lessons of the Pacific War showed that Japan had to join the West. It could not avoid the Cold War. Since Japan could not protect itself, the only solution was to have the United States do so. To this end, Yoshida was prepared to allow American military bases not only in Okinawa – an outlying island that the United States had conquered at great cost during the Pacific War – but also in Japan itself. In fact, he was ready to permit the Americans to use those bases to wage the Cold War wherever they wished, not simply in defence of Japan. In April 1950, he sent his trusted friend and Finance Minister, Ikeda Hayato, to the United States to make the offer of base rights, hopefully in exchange for a formal peace treaty and an end of the Occupation period.

The Americans were receptive and began peace negotiations in May, with American President Harry S Truman naming John Foster Dulles to lead the American team. Dulles had his own lessons of history. As a young man, he had attended the Paris Peace Conference after the First World War. Convinced that its harsh treatment of the defeated nations had led to the second global conflict, Dulles wanted to treat Japan as an equal and reintegrate it into the West. Yoshida could hardly have asked for anything more.

Unfortunately for Yoshida, the Korean War broke out in June 1950, just as his peace negotiations with Dulles were getting underway. The war pitted communist North Korea against a pro-Western regime in South Korea to determine which would rule the entire country. Although its origins were largely domestic, the Korean War quickly entangled the great powers. Within days of its start, the United States committed air, sea, and ground forces to Korea – many of them coming directly from American bases in Japan. Soon after, the Americans secured a United Nations resolution endorsing the use of force against North Korea as many other American allies dispatched their own troops to fight.

Almost overnight, Yoshida's plans for Japan were in jeopardy. The Japanese socialists bitterly criticized his allowing the United States literally to wage war from its bases in Japan. In response, Yoshida was forced to back away from his plan to allow the United States base rights after the Occupation. On the right, militant conservatives favoured a foreign policy that was pro-American but more independent. They proposed reviving the Japanese military and sending Japanese forces to Korea under the United Nations' flag. Both groups increased their criticism of Yoshida's pacifist conservativism after the Chinese entered the war actively in November.

China's entry naturally led to even stronger demands from the Japanese socialists to withdraw from the Cold War and embrace neutralism. But Yoshida was more concerned with Dulles and the Americans, who were insisting that Japan come to their aid by agreeing to create at least a fledgling military. Dulles' courting of the militant conservatives, such as Ashida and Hatoyama Ichirō, also disquieted Yoshida, as those leaders made clear their willingness to rearm Japan if Dulles helped them back into power.

Ordinarily, Yoshida could have counted on MacArthur's support. But the American general had been in Korea since the start of the war, commanding the United Nations' forces. After initial success, MacArthur had had a difficult time with the Chinese, and had called for the expansion of hostilities from Korean into Chinese territory. Truman had vetoed the idea, and had relieved MacArthur of command after the general continued to complain. MacArthur was in no position to help Yoshida against Dulles.

Yoshida also toyed with the possibility of an alliance with the Socialists. They could be counted on to stand by his resistance to Dulles' calls for rearmament. But Yoshida thought their insistence on neutralism too dangerous and their domestic programmes were odious. He was determined to link Japan to the United States as closely as possible.

As a result, Yoshida returned to his original strategy of yielding Japanese independence in foreign relations as a price for winning American military protection and economic assistance (and removing the need for serious Japanese rearmament). Indeed, he determined to yield more to the Americans rather than give in to the requirements of either the Japanese socialists or militant conservatives. He proposed that Japan sign a security treaty alongside a formal peace treaty ending the Occupation. He consented to nearly every term that the Americans, especially the American military, stipulated. Japan agreed to lease dozens of bases to the United States, but Washington incurred no legal obligation to retain them. The Americans could use the bases to defend their interests throughout East Asia, whether those interests had any bearing on Japan's defence or not. The United States, moreover, could use the bases in any way it pleased without even consulting the Japanese Government, including the deployment and even use of atomic weapons. As importantly, American soldiers would be tried by American courts under American laws.

Howls of protest greeted these two agreements, especially from the Japanese socialists and communists. Yoshida, they claimed, had returned Japan to the era of unequal treaties. Even the hated principle of extraterritoriality had been reintroduced for the benefit of the American troops. These two parties also deplored the broader consequence of these treaties: Japan was firmly aligned with the United States in the Cold War. Indeed, an angry Soviet Union and China had refused to sign the peace treaty. To add to the humiliation, the Americans had forced Yoshida to allow the rump regime of Chiang Kai-shek, which had fled to Taiwan, to sign the treaty and thus be recognized by Japan as China's legitimate government.

But Yoshida had calculated well. Although the militant conservatives chafed under the provisions of the Security Treaty, they were pleased with Japan's alignment alongside America. Perhaps as importantly, they eagerly sought the return of Japanese sovereignty so that many of the domestic reforms the Americans had carried out to liberalize Japan could be swiftly reversed, especially police and labour laws.[5] Their support ensured Diet passage of both the Security and Peace Treaties.

The end of the Occupation changed Japan's foreign policy very little. But it did touch off three years of furious manoeuvring among Yoshida's pacifist conservatives, the more militant Right, and the various groups of the Left, such as the socialists and communists. Of these, the most dynamic, and the most threatening to Yoshida, were the militants, led by Hatoyama.

These conservatives agreed with Yoshida's pro-Western orientation. But they argued that it was incomplete. They wanted Japan to be much more like West Germany: an ally, not a protectorate, of the United States. They wanted a Japan with its own military forces and its own measure of influence over its corner of the world. There would be concrete rewards for Japan, too, they argued. The United States Congress had passed the Mutual Security Act in 1953, pledging hundreds of millions of dollars toward the armament of American allies. Dulles, the new Secretary of State in Washington, was specifically calling for a Japanese military of 350,000 troops. With American dollars, Japan could build a military, and restore its regional influence, at virtually no cost to itself.

Yoshida was perfectly willing to accept this 'MSA' aid. But he balked at the one prerequisite for the militants' programme for rearmament: the revision of the 'no-war' clause of the Japanese Constitution. Yoshida understood that if he could gain the Americans' acceptance of a strictly limited programme for a new Japanese military – a military that would operate under the constraints of the 'no-war' clause – he could resist the militants' calls for constitutional revision and get American dollars. In late 1953, he sent two trusted subordinates, Ikeda and Miyazawa Kiichi, to the United States. They struck a deal with Washington to create Self-Defence Forces for Japan of 180,000 troops in exchange for American MSA funds. Significantly, Dulles told his negotiators not to press for constitutional revision. This concession permitted Yoshida to fend off the chief demand of the militant conservatives. To the infuriated socialists and communists, Yoshida blithely explained that the constitution barred war, not self-defence of purely Japanese territory.

Yoshida succeeded in preserving his basic foreign policy for Japan, but at a high political price to himself. Both the militant conservatives and business community – which were becoming more organized and effective with the end of the Occupation – were willing to overlook his intermittent coalitions with the Left to block constitutional revision so long as conservative domestic programmes were not jeopardized and so long as Yoshida maintained strong ties with the United States. But by mid-1954, conservatives of all types were dismayed that elections in 1952 and 1953 had produced no clearly dominant party. They feared that the leftist parties might unify themselves, win an election, and overthrow the conservatives' victories since the days of the Dodge Line. Since Yoshida stubbornly resisted a merger of his pacifist conservatives with Hatoyama's more militant sort, he had to go. Yoshida was out of power by the end of the year and, in 1955, a conservative union, the Liberal Democratic Party (often called the LDP), was established with Hatoyama its first leader and Prime Minister.

Hatoyama took office with an exceptionally clear set of goals. He wanted to establish Japan as an independent (yet still pro-Western) state. He wanted to abolish the 'no-war' clause of Japan's constitution and commence a substantial rearmament programme. And he sought to prevent the Socialists and their allies from ever taking power. In reality, these last two objectives were linked since the constitution could be amended only by a two-thirds' majority in any Diet vote. The leftists and enough of Yoshida's supporters (who remained as part of the new Liberal Democratic Party) would block revision unless the number of Socialist seats in the Diet could somehow be reduced quite substantially.

Hatoyama understood that constitutional revision and the weakening of the Left would require prolonged and difficult efforts. But he saw few obstacles to implementing a more independent foreign policy at once. The keystone in this programme was Hatoyama's push to normalize relations with the Soviet Union as rapidly as possible. Because the Soviets had refused to sign the San Francisco Peace Treaty of 1951, Japan and the USSR remained technically at war. If this had been the only problem, Hatoyama could have expected quick success. But there were three other problems, which were all interlinked to Hatoyama's frustration.

The first of these problems was the matter of the 'Northern Territories'. These were four islands (actually, one 'island' was itself a tiny island group) just north of Hokkaido. They had been occupied by Soviet forces in the last (and, for the USSR, the only) week of the Pacific War. Moscow claimed that the wartime Allied summit meeting at Yalta had awarded it all four islands. Every Japanese Government had pointed to the Potsdam Declaration, which, according to Tokyo, set the terms of Japan's territorial adjustments after defeat. According to Tokyo, Potsdam indicated that the four were Japanese territory. But Soviet troops were still there in 1955, and the Soviet Union had used the dispute to justify blocking Japan's admission into the United Nations.

Hatoyama saw this first problem as an opportunity. After the death of Joseph Stalin in 1953, Soviet leaders moved toward a global understanding with the West. Part of their move included a reopening of negotiations over the Northern Territories. In fact, the Soviets offered to return two of the four islands, sign a peace treaty, and permit Japan to enter the United Nations. The last of these alone would have been a terrific achievement for Hatoyama and his followers, so they leaped at the Soviet offer.

That leap brought into play Hatoyama's two other problems: Yoshida's pacifist conservatives, and the United States. In part, Yoshida's followers were concerned that Hatoyama would harm relations with Washington. But it is just as true that they were willing to fan American fears of an independent Japan in order to prevent Hatoyama from enjoying political success at home. Since Yoshida still had a stronghold within Japan's Foreign Ministry, his group was well placed to fan those fears and delay progress in the negotiations. In this way, rivalries within Japan's conservatives became entangled with the wider Cold War between the United States and Soviet Union.

The power of those rivalries was revealed in the path of the Northern Territories negotiations. Japan's Foreign Ministry, under Yoshida's control, rejected the Soviet offer of two of the four islands. The Soviets then proposed that a peace treaty be signed that left all territorial issues for the future, a formula they had successfully used with West Germany in late 1955. Since this would still have represented a great achievement for Hatoyama, the Foreign Ministry blocked this proposal, too. Moscow then applied pressure by closing some northern Pacific waters to Japanese fishing boats – which allowed Hatoyama to bypass Yoshida's Foreign Ministry and use his own man, Minister of Agriculture and Forestry Kōno Ichirō, to deal with the fishing dispute. To no one's surprise, Kōno returned with an agreement allowing fishing – once Japan had concluded a general peace treaty with Moscow. Hatoyama vowed he would go to Moscow to obtain a treaty personally and would retire as Prime Minister once he had succeeded.

Hatoyama had managed to overcome Yoshida's allies within Japan. But he could not prevail against Yoshida's friends in the United States. As Hatoyama prepared for his trip, in the summer of 1956, Dulles announced that if Japan agreed to cede two of the northern islands to the Soviets, the United States might permanently occupy Okinawa. The gain of two small northern territories would hardly offset the loss of any chance to regain Okinawa, as Hatoyama fully understood. He, therefore, bitterly agreed to the West German formula: Japan would sign a peace treaty. It would enter the United Nations by the end of the year. But it would not resolve the territorial issue. And Hatoyama would still have to resign. The two wings of the conservative party had fought to a deadlock.

Their fight also ensured that Hatoyama would not achieve his other goals. Distracted by the dispute over the Moscow negotiations, Hatoyama was unable to gain the Diet's approval for an electoral redistricting plan that would have harmed the Socialists. Their strength, combined with Yoshida's, in turn ensured that Hatoyama's push for revision of the constitution's 'no-war' clause would come to nothing. The fight may have been a tie, but the larger victory was clearly Yoshida's. Japan would remain pro-Western, but thoroughly pacifist.

Yoshida's grand strategy also survived its next major challenge: the revision of Japan's security treaty with the United States. The challenger was Kishi Nobusuke, who became Prime Minister two months after Hatoyama stepped down. Kishi's background was, if anything, even more colourful than Yoshida's. He had been a superior student at Tokyo Imperial University, but had gone into the Ministry of Commerce and Industry, not the Foreign Ministry, upon graduation. Kishi swiftly joined forces with the Imperial Army's economic reformists of the 1930s, and rose to become Minister of Munitions (the wartime title of Commerce and Industry) under Tōjō. After years of political exile during the American Occupation, he became an active politician in 1953. He was a militant conservative, and his first duty had been to lead a commission charged with revising the postwar constitution to overthrow, among other things, the 'no-war' clause.

Kishi deeply believed in his own personal responsibility for Japan's sufferings during the Pacific War. He had no intention of challenging

Japan's connection with the United States. But he shared Hatoyama's conviction that Japan should not be a supine American protectorate, and he thought the Security Treaty represented a humiliating retreat to semi-colonial status. Accordingly, he set out to revise that treaty (as a step toward eventual revision of the constitution and Japanese rearmament). Kishi also wanted to normalize relations with China, hopefully with fewer complications than Hatoyama had encountered in his efforts with the Soviet Union.

Interestingly, the Japanese Socialist Party strongly supported Kishi's efforts to revise the Security Treaty and restore relations with China. The Socialists resented Japan's alignment with America and seized upon incidents, such as an American soldier shooting a Japanese woman scavenging on a base's firing range, to highlight the cost of Japan's participation in the Cold War. The Taiwan Straits Crisis of 1958 – which threatened a shooting war between Taiwan and China and possible American involvement – demonstrated to the Socialists that that cost could become exceptionally dear. The crisis also neatly illustrated the wisdom of Kishi's programme to normalize relations with the Chinese.

To Kishi, growing tensions in the Pacific made Japan all the more valuable to the United States and all the more worthy of American concessions. As early as mid-1957 he had personally toured America and had received a standing ovation after an address to the United States Congress. The reason for the applause was not difficult to see. Kishi had expanded Japan's Self-Defence Forces. He had travelled to nations throughout the Southwest Pacific and had assured them and the Americans of Japan's willingness to take part in collective security arrangements against the communists. Washington was especially pleased by his trip to Taiwan.

In exchange, Kishi won much from the United States. Japan was allowed to provide developmental funding for Okinawa, an entering wedge for the eventual reassertion of complete Japanese sovereignty there. American influence won loans to Japan from the International Monetary Fund and similar agencies. Most importantly, the United States agreed to consider revising the Security Treaty.

But Kishi lost, too. His trip to Taiwan infuriated China and alienated the Japanese socialists. Even then, Kishi might have enjoyed their support for the Security Treaty's revision and opened the way for his grander foreign policy programme. But he irretrievably damaged these prospects by revealing his arch-conservative views in two political struggles at home. In the first, Kishi sought to greatly strengthen the powers of the police, provoking bitter strikes from organized labour, whose leaders (correctly) saw themselves as Kishi's target. In the second, Kishi struck directly at one of the most influential unions in attempting to implement a new set of regulations for teachers. Because labour formed the backbone of the Japanese Socialist Party, Kishi's domestic initiatives – which both failed – ended all hope of Socialist support for Kishi in any security revision fight.

In fact, it was even unclear whether Kishi could count on much of the LDP to support him. Yoshida's wing already suspected that Kishi meant to rewrite the treaty to permit, or even require, the rearming of Japan and its participation in a Pacific version of America's North Atlantic Treaty Organization (NATO) military alliance. The Taiwan Straits Crisis of 1958 led other party leaders to fear that under any new treaty Japan might be obligated to join American efforts to defend Taiwan, efforts that could lead Japan again to war with China.

Kishi, therefore, was obliged to compromise. In early 1959, he announced that the United States had dropped its insistence that Japan amend its constitution to allow substantial rearmament (meaning Kishi had given up his plan for fairly swift revision). And he decided that the revised treaty would not oblige Japan to defend any territory outside the home islands, not even Okinawa. This step displeased his own followers, who wanted Japan to reassert control over Okinawa in any way possible. Indeed, the leaders of the militant factions of the LDP were convinced that Kishi had betrayed their hopes for a fundamental reordering of Japan's foreign policy. They took the unusual and, for Kishi, insulting step of conferring directly with the American Ambassador to include Okinawa within a limited Japanese defensive umbrella. Kishi responded by ejecting these leaders from his cabinet and allying with the Yoshida factions in July 1959.

Kishi's retreat also caused problems with the Americans. Washington favoured Japanese rearmament and participation in collective security efforts throughout East Asia. It successfully insisted that the new treaty specifically permit American forces operating from Japanese bases to protect 'the Far East', not only Japanese territory. The Socialists immediately demanded clarification: could this involve Japan in war with China if, for example, American planes bombed Peking from Japanese bases? Kishi's Foreign Minister [6] gave a confusing, fumbling response that encouraged the Socialists to go forward with plans for a mass protest movement against the new security treaty.

Otherwise, the United States was quite accommodating in the treaty negotiations. The new agreement bound Washington to defend Japan, an obligation that the 1951 accord had not included. The United States agreed that Okinawa and nearby islands were under the 'residual sovereignty' of Japan. Japan was relieved of much of its financial obligation to support the bases within its territory and was not required to increase its Self-Defence Forces. American troops would no longer have the right to act against internal instability within Japan. It was a significant achievement for Japan and Kishi, and Yoshida's pacifist conservatives could accept it as well.

The ratification of the Security Treaty of 1960, however, was pure Kishi, which is to say it was exceptionally maladroit. Kishi was also exceptionally unlucky. The Socialists, who began their attack by showing pictures of Kishi as a member of Tōjō's wartime cabinet, claimed that the treaty violated the 'no-war' clause of Japan's constitution since it recognized Japan's right to armed self-defence. They returned to the treaty's 'Far East' provision, asking how deeply this might involve Japan if the Cold War turned hot. Kishi's reassurances on this score were literally shot down when, on 1 May, the Soviet Union destroyed an American U-2 spy plane over Soviet territory. It was an open secret that U-2s operated from bases in Japan, and Moscow's threat to fire nuclear rockets against bases used to violate Soviet airspace made Kishi's position difficult.

Ordinarily, Kishi might have waited for the U-2 affair to blow over. But his government had arranged for American President Eisenhower to visit Japan in mid-June. Kishi badly wanted the Diet's ratification in hand by then. Knowing this, the Socialists – who could not defeat the treaty in a party vote – planned to delay ratification to embarrass Kishi. Their plan worked very well. The Diet session expired on 26 May with the treaty unratified. Kishi then called for an extension of the session. The Socialists physically tried to block the Diet's Speaker from calling for a vote on extension and were forcibly removed from the Diet by police, whereupon the Liberal Democratic Party extended the session and ratified the treaty.

There were fierce howls of protest over Kishi's strongarm tactic, and in the mass demonstrations that followed one student was crushed to death as protestors and police battled with rocks, clubs and tear gas outside the Diet building. But Kishi could not stop: he wanted approval by the Diet's upper house. There, his earlier clash with the right-wing members of his own party undid him. The upper house would ratify the treaty, but only if Kishi then resigned as Prime Minister. Kishi had little choice but to comply. He also was compelled to cancel Eisenhower's visit, as the Americans expressed concern over the President's safety in light of the turmoil in Tokyo. It was an ignominious end to Kishi's colourful career in government.

Ironically, it was also an end to Japan's period of ignominy in foreign relations since the end of the Second World War. The Security Treaty of 1960 ended the unequal status of the 1951 agreement and made Japan an ally of the United States (albeit an ally without a significant military component), not merely a protectorate. That treaty also inaugurated a decade that would see Japan leap from the status of a 'third world' nation toward one with economic muscle second only to the United States. That muscle would inaugurate a new era in which trade issues came to dominate Japan's foreign relations.

Notes

[1] Not coincidentally, Burma and the Philippines were poorest in 'national defence resources'.

[2] That is, advisers to the Emperor.

[3] The peace group often argued that for the army to continue the war too long invited a communist takeover of Japan. After Japan's surrender, members of this group used much of the same argument with the Americans as they attempted to curb the Occupation's internal reforms. How great was the threat of Japan 'going communist'? This question is difficult to answer. The police apparatus of wartime Japan operated quite efficiently and the Japanese Communist Party was small. But the conservatives' fear was great, and it was fed by a certainly accurate reading of a Japanese populace that was at its wit's end by the summer of 1945. The swift organization of labour, and its militance, early in the Occupation make clear that fundamental reordering in Japan was certainly possible, but not whether it was likely.

[4] Yoshida's party, the old *Seiyūkai*, rose from 131 to 264 seats. Ashida's *Minseitō* fell from 132 to 69, and the Socialists from 144 to 48.

[5] Another, related factor was the militant conservatives' desire to swiftly end the Occupation so that a number of their leaders, who had been 'purged' by the Americans, could be allowed back into active politics.

[6] Fujiyama Aiichiro.

8
The quiet giant

After the Security Treaty Crisis of 1960, Japan's leaders never again challenged Yoshida's formula for Japanese foreign policy. Japan would not build a strong military to threaten its conservative leadership and sap its few resources and precious national wealth. The nation's protection would be served by an alliance with the Americans, at the price of less than full independence in Japan's foreign relations with the rest of the world. Indeed, in every major and even most minor matters of international relations, Japan would follow the lead of the United States. The nation's prosperity, however, would be promoted by positive action by the Japanese Government. As importantly, the Americans, keenly aware in retrospect of how close they had come to damaging their alliance with Tokyo, largely ceased their pressure for a larger military role for Japan. Washington also actively aided Japan's entrance into the community of the wealthy nations, symbolized by Tokyo joining the OECD in 1964 over some European objections. By the 1960s, the first, dramatic results of Japan's drive to prosperity were evident. By the 1970s and 1980s, Japan's economic success was so great that it threatened to overwhelm its good relations with the United States and the rest of the West, and return Japan to its nightmare: international isolation. At the same time, a worsening of the Cold War between the West and the Soviet bloc renewed American pressures on Japan to rearm. As a result, contemporary Japanese foreign policy has devoted itself to one overarching issue: how to reconcile its economic gianthood with its determination to avoid all questions of politics in international affairs or, to put it less charitably, how to remain a political dwarf. Ironically, the end of the Cold War with the arrival of the 1990s has only made that reconciliation more difficult.

Even as Kishi Nobusuke secured final ratification of the Security Treaty of 1960 and resigned, Japan had made remarkable economic progress. In that year its income per capita exceeded US$350 per year. This sum may not seem impressive, as it was approximately equivalent to Mexico's, but it represented the highest figure for any Asian nation. Much more illuminating was the rate of Japan's economic growth: over 15 per cent per year in 1960 alone. Japan's ability to sustain that pace explains its swift rise to economic pre-eminence, second only to the United States, in the three decades that followed.

What were the keys to Japan's performance? Many critics from the 1970s to the 1990s argued that Japan employed 'neo-mercantilist' tactics that are reminiscent of the old commercial rivalries of the European empires of the seventeenth and eighteenth centuries. In essence, each empire sought to deprive its rivals of riches by denying them trading privileges with itself and its colonies while seeking to break into trade with its rivals' holdings. Japan, its critics argue, has sought to deny the West free access to the Japanese market while taking advantage of the free access that the West, and especially the United States, permitted to Japan.

There is a measure of truth in these charges. But they are too simplistic and, as important, wrongly timed. There can be little doubt that Japan was highly protectionist during and immediately after the Occupation. Measures such as the Foreign Exchange and Foreign Trade Control Law (1949) and Foreign Investment Law (1950) gave sweeping powers to Japan's Ministry of International Trade and Industry (MITI). MITI used these powers radically to restrict imports into Japan and block foreign control of Japanese firms. Indeed, since MITI had virtual control of access to foreign goods, technology, and capital, it could also use these powers to provide quite strong guidance to its vision of which Japanese industries ought to be developed. Overall, MITI's guidance worked well, as it sought to protect and develop Japanese enterprises in the textile, petrochemical, steel, shipbuilding and automobile industries.

Yet MITI was by no means the only reason for Japan's early successes. The American-imposed 'Dodge Line' forced a radical austerity programme upon the Japanese Government and a correspondingly radical shortage of capital for Japanese businesses at the end of the 1940s. One result was the rising influence of large banks, which became the centres of *keiretsu*, large alliances of companies that came to dominate many key sectors of the economy. Because these sectors were controlled by relatively few companies, it was easier for MITI, in turn, to provide guidance. Because Japanese banks became quite influential, it was likewise possible for them, often in conjunction with MITI and the Ministry of Finance, to form a consensus concerning what that guidance should be.

Japan's remarkable political stability also eased the way to achieving consensus between government and business. The Liberal Democratic Party monopolized Japan's cabinets from its formation in 1955 until its disruption in 1993. To a remarkable degree, LDP politicians, ministry bureaucrats, and business leaders developed a closeness of background and purpose not matched outside of Japan. These men frequently were personal friends of long standing, having attended the same schools and universities and, at times, having married into each other's families.

Those schools and universities also have contributed powerfully to Japan's economic success. Japan's school years and school days are considerably longer than American or even European ones. The curriculum is more focused, with fewer electives, and the examination system is renowned for its exceptionally exacting standards.

Even so, none of these should be taken as an indication of Japan's uniqueness, cultural or otherwise. In fact, nearly all of the factors that have contributed to Japan's sparkling economic performance have arisen from its foreign-policy considerations. The laws that empowered MITI were Occupation-era laws, which the Americans not only permitted but actively encouraged. In large part, American consideration for Japan, and American encouragement of Japanese protectionism, arose with the coming of the Cold War to East Asia. With the fall of China to communists in 1949, American 'Cold Warriors' such as George F. Kennan emphasized that the West needed a strong Japan as its Asian bastion. Kennan was convinced (correctly, as it turned out) that the Cold War against the Soviet bloc would be a protracted affair ultimately settled by economic, especially industrial, strength. Yet his tour of Japan convinced him that the Japanese economy, far from being a source of that strength, was in such desperate straits that Japan might become a permanent ward of the United States, requiring years and perhaps decades of economic assistance, instead of providing that assistance. Kennan and other American leaders, therefore, supported steps that would enable Japan to revive its economy. MITI's power, and the banks' formation of the *keiretsu* under Dodge's line, owed much of their creation to American concerns.

As well, the Americans were convinced that the best way to avoid a Third World War was to avoid the economic nationalism, even autarky, that had prevailed in the 1930s. As a result, the United States led an effort to forge an international economic and financial regime that would further international commerce, and hence prosperity, and prevent relapses into economic nationalism by creating (and largely capitalizing) international financial institutions such as the World Bank and the International Monetary Fund (IMF). Member states, who came to form their own sort of club, termed the Organization for Economic Cooperation and Development (OECD), promised to abide by a set of rules prohibiting undue protection of their domestic economies from either foreign competition or investment. Another accord, the General Agreement on Tariffs and Trade (GATT), established many of these rules.

Immediately after the Occupation ended, Japan was regarded as a developing country, not too different from Mexico or Greece. Its economy was relatively small, it took a minor share of global trade, and its prospects for prosperity appeared problematic. The Korean War boom helped commence the strong growth of the Japanese economy through the 1950s. At the same time, Japan received funding from the World Bank, becoming a member through Washington's sponsorship upon regaining its independence, and enjoyed Western (or at least American) tolerance of its protection of its domestic industries. Also, the Americans had generously allowed the Japanese unit of currency, the yen, to be pegged (that is, its exchange rate set) at 360 yen to 1 dollar. The effect was to 'undervalue' the yen relative to its actual market value, so that goods produced in Japan would be cheaper in foreign markets. In other words, the Americans were encouraging the growth of Japanese exports through their currency policy. At the same time, Washington raised no criticisms of MITI's protectionist methods. The top objective was a prosperous Japan joining the West as a full-fledged member of its economic order (and Cold War alliance) as soon as possible.[1]

Japan became a member nation of GATT in 1955 over European objections that MITI's protectionism was inconsistent with GATT rules. Only strong American support allowed Japan's bid to succeed, and then it required American assurances that Japanese protection would end. But the Japanese also understood that they would have to adhere to GATT rules and began by gradually reducing and eliminating tariffs and other trade barriers. Indeed, Japanese business actually rebelled against a MITI plan in 1963 to further assist 'designated industries' by subsidies and other assistance. A similar liberalization of MITI's restrictions on foreign exchange followed Japan's joining the IMF as a leading member (that is, under Article 8) in 1964. Thereafter, Japan became a lender of funds to the IMF and World Bank to help other nations prosper. Japan helped found the Asian Development Bank in 1966.

To the surprise of some MITI bureaucrats, LDP politicians, and business leaders, Japan not only survived the lifting of tariff and foreign exchange controls, it prospered quite well. It was no surprise, however, to Prime Minister Ikeda Hayato, who literally travelled the world boosting Japanese products. A disdainful Charles de Gaulle of France called Ikeda a 'transistor radio salesman', but Ikeda got results, including the establishment of an American–Japanese committee for trade and economic affairs that would meet annually. And his Income-Doubling Plan – to see Japan's income double from 1960 to 1970 – actually understated the results. During the 1960s, telephones, washing machines, television sets and automobiles, which were rare, luxury items before that decade, suddenly became commonplace. Even the people's diet – and physiques – changed.

Less happily, other changes also came. The explosion of Japanese exports came just when the United States was showing the first signs of the strain of bearing the chief burden of the Cold War for the West. America's trade surplus (the value its exports exceeding that of its imports) slipped to a trade deficit by the early 1960s. Specific American industries were surprised at the competitiveness of their Japanese rivals. Presuming that these rivals had been unfairly protected by the Japanese Government, the Americans began to clamour for their own protection.

Among the first signs of American–Japanese trade friction was the declaration by American textile union members in early 1961 that they would refuse to work with cloth imported from Japan. The immediate issue concerned Japanese imports of mens' and boys' suits. Although these were a tiny fraction of the total suits sold in America, the union cited the huge wage differential – American textile workers were paid $2 per hour, the Japanese 14 cents – to argue that the trickle would soon become a floodtide.

American President John F. Kennedy persuaded the textile workers to cancel their work boycott, but he did open negotiations to avert that floodtide of imports from Japan. Ironically, the floodtide had been made possible by the Japanese Government's ending restrictions on imports of raw cotton and wool in order to meet GATT requirements. Japanese companies had increased their imports of these items, and hence exports of finished goods, such as suits, enormously. By the autumn of 1962, Japan had agreed reluctantly to quotas on its exports of textiles to the United States within the framework of a comprehensive, international Long Term Arrangement covering textiles.

A second disappointment soon followed. To address the growing American trade deficit, in the summer of 1963 the American Congress enacted a tax on interest collected from American investments abroad. Canada, fearing that the tax would cause dollars to leave its country, protested and received an exemption. Japan, eager for more capital, also sought exemption. In a sign that the United States was no longer viewing Japan as a developing nation in danger from internal, communist subversion, but rather a new economic power that actually had interests that could injure American workers, Washington refused, considering its economic troubles greater than the need to accommodate Tokyo for the sake of the military alliance.

Even so, that alliance remained rock steady. The Cuban Missile Crisis of October 1962 brought the world to the brink of nuclear war. Many Japanese, including members of Ikeda's cabinet, felt that Japan should reconsider its ties with the United States, arguing that American bases in

Japan would act as nuclear lightning rods for Soviet missiles. The Prime Minister stood by the President. As well, American involvement in Vietnam generated widespread criticism and disillusionment in Japan. Nightly newsreels of American helicopters assaulting fleeing Asian soldiers, and American patrols through Asian villages, stirred uneasy memories of the worst days of the Pacific War. Still, Ikeda and his successor, Satō Eisaku, refused to rebuke Washington for its intervention. In a controversy that many Japanese saw as just as significant for Japan's place in the West but not of it, Ikeda and Satō stood by the United States' constant opposition to permitting the communist People's Republic of China to take China's seat (then still held by the Nationalist regime, which had fled to the island of Taiwan in 1949) in the United Nations. Membership in GATT and the OECD was more important to 'salesmen' Ikeda and Satō than political gestures to fellow Asians on the wrong side of the Cold War in Asia, especially as Japan's foreign trade increasingly rotated around the industrialized countries of the West, instead of its prewar emphasis on nearby East Asia.

But there was one issue concerning fellow Japanese that increasingly dominated Japanese foreign policy by the end of the 1960s: Okinawa. Technically, Okinawa did not belong to the Japanese archipelago, but nearly all Japanese regarded it as part of their home territory. At the start of the Occupation in 1945, the Americans had thought differently. Okinawa had been the gateway to an eventual invasion of Japan and American air and naval bases there would be important in restraining any revival of Japanese militarism. When the Cold War came to Asia by 1949, these bases acquired even greater value, only now their targets would be Soviet or Chinese.

By the same token, however, the Americans valued alliance with Japan much more by 1949. That value was proven when war in Korea broke out a year later. As a result, the San Francisco Peace Treaty allowed Japan to assert a 'residual sovereignty' over Okinawa, although actual possession and control remained firmly in American hands.

This compromise became increasingly awkward over time. By 1957, the Americans were placing Intermediate Range Ballistic Missiles (IRBMs) with nuclear warheads on Okinawa, just in time for the second Taiwan Straits crisis a year later. That crisis, begun over a struggle between Taiwan and communist China for control of two small islands in the straits, threatened to escalate into a full-blown war, possibly nuclear and, therefore, possibly involving Okinawa. Nearly all Japanese, including most members of the ruling Liberal Democratic Party, believed that Japan had no fundamental quarrel with China. They were uncomfortable with the thought of being dragged in to a confrontation through the Okinawan bases. Washington heard Tokyo's concerns, but the

American military pointed out that the straits crisis proved why the United States should not change Okinawa's status.

Ironically, the first impetus toward Okinawa's reversion came from Prime Minister Kishi Nobusuke's manoeuvrings in revising the United States–Japan Security Treaty in 1960. Kishi hoped to have the new treaty include Okinawa within its defence perimeter. Because Okinawa was American territory, its inclusion would obligate Japanese forces to defend part of the United States, making the treaty truly one of mutual defence instead of the one-way 1951 agreement which offended Kishi's sense of Japanese sovereignty.

Like so many of Kishi's other initiatives, this one was howled down by treaty opponents and became part of the acute treaty crisis of 1960. But the episode highlighted for the Americans how sensitive their continued retention of Okinawa was. President Kennedy's new ambassador to Japan, Edwin O. Reischauer, privately warned that sooner or later the issue would poison relations. A sympathetic Kennedy issued a communiqué in 1962 clarifying that Okinawa was essentially Japanese territory and would be returned eventually. The United States began to reduce its economic assistance to Okinawa as Japan was allowed to increase its aid, an important symbolic development. And, in August 1965, Prime Minister Satō became the first Japanese leader to visit Okinawa since the Pacific War.

The timing of Satō's visit was spoiled, however, by that one constant in the Okinawan issue: its continued military value to the United States. Just before his arrival, the Americans decided to base their long-range, strategic bombers, B-52s, on Okinawa. And their involvement in the Vietnam War was rapidly escalating, with some American troops bound for Vietnam deploying out of Okinawa.

In the end, though, America's debacle in Vietnam eased the way to Okinawa's return to Japan. Satō's quiet but firm support for American involvement in the war, in contrast to some European criticisms, impressed Washington, especially after a (non-nuclear) bomb-laden B-52 crashed on Okinawa in November 1968. More to the point, that same month Americans elected Richard M. Nixon President, and Nixon knew he had to get the United States out of Vietnam. This position encouraged progress in returning Okinawa to Japan for two reasons. America would have less direct use for bases there in supporting its involvement in Vietnam. More important, the Nixon Administration realized that its pull-out from Vietnam would make strong allies elsewhere in Asia all the more necessary for the United States. Already some members of Japan's ruling Liberal Democratic Party were arguing that Japan had to return to Asia and Asianism with the decline of American power meaning, among other things, much closer relations with communist China.

Satō, a strong and steadfast friend of America, attempted to defuse these criticisms by launching a major, public drive for Okinawa's reversion. He saw Nixon as a hardline Cold Warrior, and, therefore, able to fend off criticisms from America's right-wing which opposed surrendering any American military installations anywhere. Satō was right, but only to a point. Nixon was prepared to relinquish Okinawa, but not American bases there. Ordinarily, this stance would have presented no problems to Satō; there were American bases in Japan proper, after all. But under the 1960 Security Treaty there was a crucial difference: the Americans had committed themselves to consulting with Japan before introducing nuclear weapons into bases in Japan, and both countries operated on the assumption that nuclear weapons had never been stationed at bases there. But it was obvious that the Americans did have these weapons on Okinawa. Nixon insisted that they be allowed to remain, putting Satō in a difficult position.

If Satō had decided to press Nixon hard, he may well have won his point. Secretly, the Americans had determined that their growing Intercontinental Ballistic Missile (ICBM) force and their rapid increase of nuclear missiles aboard nearly invulnerable submarines made deployment of nuclear weapons on Okinawa unnecessary. Nixon, however, used the nuclear issue as a bargaining chip, and traded well for it in an agreement with Satō in November 1969. In exchange for an understanding that the Okinawan bases would operate under the same procedures as bases in Japan regarding nuclear weapons, Japan agreed to 'support' American foreign policy in South Korea, Taiwan (a point which angered the 'Asianist' wing of the LDP desiring closer relations with communist China), and the 'Far East' in general. Most observers viewed these provisions as American prodding to have Japan take a more active role in the defence of Western interests in East Asia. Indeed, the agreement specifically bound Japan to defend Okinawan territory. Even so, Satō won his main point: Nixon promised to return Okinawa by 1972.

Unhappily for Satō, by that time the Okinawan issue had been entangled with two other major problems in Japan's diplomacy: a new trade crisis and Nixon's electrifying turnaround over communist China.

The new trade crisis revolved around an old problem: textiles. Japan had accepted the Long Term Arrangement of 1962 limiting its sales to fixed percentages in the United States and Europe in the hopes that their markets would grow and, with that growth, permit larger absolute Japanese sales. In addition, this seemed to be the only way for Japanese textile manufacturers to gain any European sales, as the European Economic Community, or 'Common Market', had been rigidly protectionist in its treatment of Japan.

These foreign markets did grow, hugely, throughout the prosperous 1960s. But Japanese sales grew even faster, because the Long Term Arrangement had become obsolete through technology. The popular textiles since the 1960s have not been cotton or wool, but new fibres such as nylon and polyesters. Thanks in part to MITI's aid to Japan's petrochemical industry, the materials for making these new materials were available and cheap for Japanese textile manufacturers. The result was a new surge of Japanese textile exports, and new cries for protectionism. The most important of these cries came from American companies and unions.

Most of these companies, and their workers, were in America's southern states. Those states played a key role in Richard Nixon's presidential campaign of 1968. After his inauguration the following year, he felt obligated to deliver on his campaign promises of protection for American textiles. As well, the Nixon Administration began what would become a new tradition in American diplomatic dealings with Japan: a prominent member of the American business community[2] would publicly unleash a general attack on Japanese trade barriers. Satō's government was surprised and unprepared to respond: it had thought that Japan had an outstanding record of liberalizing its trade. A concerned Satō agreed to direct, bilateral negotiations with Nixon, although his advisers thought the issue would have been better settled through the multilateral GATT mechanisms.

Satō might have thought so privately, but he was unprepared to disturb the bedrock of Japan's postwar diplomacy: the American alliance. Besides, Okinawa had not yet been returned, and he did not wish to allow Nixon any pretext for deciding to keep it. At his meeting with the American President in November 1969, Satō agreed to a secret pact substantially restricting Japanese textile exports to the United States. Nixon reaffirmed that Okinawa would be returned, and that American bases there would operate under the same rules as those in Japan.

Satō needed a secret deal because Japanese elections were scheduled for December and he had no desire to injure his standing by revealing what was sure to be an unpopular agreement. Once the elections passed, however, the Americans insisted that Satō make the terms of the pact public. When Satō reviewed what the Americans thought those terms were, he balked and insisted that the Americans file their complaints under GATT procedures, outraging Washington and leading American textile representatives to speak of a 'Pearl Harbor in the clothing industry',[3] as they began to usher an import quota bill through Congress.

The Japanese textile industry was equally angry. Its spokesmen argued that the United States was betraying the system of free trade it had created and championed. Japanese newspapers seemed rather pleased at the impasse, declaring with evident pride that Japan at last had stood up to the Americans and that finally the Occupation mentality was over.

The textile crisis, however, was not over. Nixon, furious at the latest turn of events, had no compunctions about surprising Japan twice in the summer of 1971 in what became known as the 'Nixon Shocks'. In July, Richard Nixon, the Cold Warrior who had made a political career out of anti-communism, especially the Asian kind, announced his intention to visit communist China and begin to normalize relations between Beijing and Washington. Satō, long-time friend of the United States and leader of America's top ally in Asia, received only hours' advance notice.

Equally profound was a second shock, delivered in August: the United States ended the dollar's convertibility to gold. Henceforth, the dollar's value would no longer remain fixed relative to any other nation's currency. Nearly everyone predicted that the yen's value would rise against the dollar, making Japanese exports to the United States more expensive and hence less competitive. Washington also imposed a 10 per cent surcharge on all imports into the United States to make Japan's position even worse. The effects on the world's trade and finance were immediate and stupendous.

Satō was mortified by Nixon's crude treatment. But his faith in the centrality of the American alliance to Japan's diplomacy was unshaken. He immediately aligned himself with the American line of a 'two-China' policy. That is, communist China would be allowed into the United Nations, but Taiwan should not be removed from UN membership. This stance pleased the Americans, but alienated leaders of the left-wing political parties in Japan, who blasted Satō for submitting to Washington's bullying. For Satō, it was more worrying that influential leaders within his own Liberal Democratic Party were sounding the same complaints. One of these, Tanaka Kakuei, was bold enough to lead a successful rebellion against Satō and replace him in June 1972.

Even so, Tanaka was not interested in disturbing the American alliance. He immediately journeyed to Honolulu to secure Nixon's consent for Japan to re-establish normal relations with communist China even more rapidly than America. Nixon agreed, but Japan was to buy over US$700 million worth of American goods to ease its trade imbalance with the United States. And Tanaka's primary concern when he arrived in Beijing in September was whether communist China would accept Japan's Security Treaty with the United States as an unalterable fact of the Cold War order.

The Chinese understood quite well that Japan would not end the American alliance. They were prepared to accept its continuation. But they made clear that Japan would have to jettison any formal recognition of Taiwan as a separate entity: the peace treaty between Japan and

Taiwan that Yoshida had signed over twenty years earlier would have to go. Tanaka was agreeable and Taiwan was out in the cold, a rather abrupt swing compared to Satō's Chinese policy.

For just that reason, Tanaka's swift pace of re-establishing relations with communist China was criticized from many sides. Obviously, Taiwan was outraged. This also meant that some LDP leaders, with strong ties (both political and economic) to Taiwan, were angered. Even those leaders who favoured closer ties with China, though, were unhappy with China's dismay at Tanaka's formal speech in Beijing. Tanaka, making the first official reference since 1945 to the Second World War in Asia, merely apologized for the 'great inconveniences' (*tadai no gomeiwaku*) Japan had caused China, a phrase that, to Chinese ears, hardly seemed adequate for eight years of harsh occupation and millions dead. Last but hardly least, the Soviet Union bristled at Tanaka's joining the Chinese in denouncing the USSR's attempts to achieve hegemony in East Asia.

Tanaka did pay attention to the Soviet Union, but only cursorily. His trip to Moscow in September 1973 was a largely symbolic affair. Nothing was accomplished concerning the still top issue for Japan: the Northern Territories. Bold Japanese ideas (not fully shared by Tanaka) to exploit the resources of Soviet Siberia, such as the oilfields at Tyumen, came to little in the face of Soviet stalling, Chinese opposition,[4] and Japanese indecision.

For all this, the focus of Japan's foreign policy throughout the 1970s remained on the United States and on economic issues, especially Japan's growing trade surplus with America. This surplus had helped trigger the second Nixon shock. Tanaka lost no time in agreeing to export quotas in textiles, explaining to angered Japanese manufacturers that American bullying left him no alternative. The bully returned in early 1973 as the United States devalued the dollar,[5] making Japanese exports to American consumers more expensive. The Japanese had sought European allies to resist Washington's pressure. Not a single country sided with Tokyo, leading some to recall Japan's diplomatic isolation of the 1930s. An American embargo on soybean exports that summer (Japan was a leading consumer) only increased the sense of being alone. Tanaka grew concerned that the new value of the yen, and Japan's image of a predatory merchant nation, would seriously damage the Japanese economy. Indeed, he began a very substantial domestic spending programme to build industrial cities beyond the main island of Honshu, to impose controls to improve the environment and, in general, to improve the ordinary Japanese citizen's standard of living. In part, of course, this was pork-barrel politics to enhance Tanaka's popularity. But it was also Tanaka's desire to increase domestic prosperity and spending to eliminate, or at least reduce, Japan's burgeoning trade surplus with the West.

Tanaka's logic was clear: a richer Japan would buy more imported goods, redressing the surplus.

The plan did not succeed because, in fact, Japan did not become more prosperous in the early 1970s. But prosperity was injured, in 1973, not by the yen's high value or the soybean embargo or wider Western resentment against Japan's trade surplus. Instead, the outbreak of the October War in the Middle East between Israel and a coalition of Arab nations shook Japan's – in fact, much of the world's – economy to the core. Even the American alliance felt the tremors. The reason was oil. For the first time, the Arab oil-producing states employed their growing share of global oil production as a diplomatic weapon. The Arabs asked for Japanese support against Israel. Certain that Washington would disapprove, Tokyo refused to align with the Arabs, who immediately reduced their exports to Japan. This action created a crisis, since Japan relied upon oil for most of its energy needs, and upon the Arabs for most of its oil.

Even then, Japan refused to act without an American go-ahead. As one Western European nation after another bowed to Arab demands, Japan waited until American Secretary of State Henry Kissinger arrived in Tokyo as he returned to America from the Middle East. Kissinger hoped to keep Japan aligned with Israel, but the Japanese argued strongly that their economy would not survive a protracted oil embargo. Once Kissinger relented, Japan publicly supported Arab demands in the war, cut off fuel supplies to American jets based in Japan, announced an aid package for Palestinian (Arab) refugees, and plans for investment in the Middle East. This last item included investment in oil production facilities. During the oil crisis, Japan had been disappointed by the lack of sympathy the (usually) American-owned oil companies had shown to Japan's unusual vulnerability to an oil cut-off. One way to reduce that vulnerability was to begin partnerships with Arab oil producers.

By the end of 1973, the American alliance no longer looked unshakable. The Nixon shocks and subsequent bullying over the yen's value and additional restraints on Japanese exports, the soybean embargo, and the lack of American support during the oil crisis, all induced doubts about Washington's reliability. The American pull-out from South Vietnam, and that country's collapse in 1975, only deepened such doubts, as did the lacklustre performance of the American economy after the global recession of 1973–4.

Japan's economy, in contrast, recovered more quickly and more strongly than any other industrial nation's. The key was controlling price inflation. In part, a reinvigorated MITI launched an ambitious programme to reduce Japan's energy use (and hence energy costs) and especially Japan's reliance upon oil. Just as importantly, the government fought inflation through the well-tested means of raising interest rates and shrinking the money supply. These steps hurt Japanese consumers, but helped the economy in the long run. Western countries, including the United States, concentrated on lowering unemployment, or reducing its impact, through increased social spending which further raised inflation. Prices of Western goods rose, making Japanese products – and exports – more attractive to Western consumers.

As a result, Japan's trade surplus with the other industrial nations increased even more through the 1970s, despite the yen's high value. Western complaints against Tokyo increased, too, promising further friction over trade and currency issues. Indeed, it is fair to say that the central story of Japanese diplomacy since at least the mid-1970s is a story of the diplomacy of trade and currency and Japan's increasingly important role in setting the tone and pace of trade and currency negotiations.

One example of this setting was the Tokyo Round of negotiations under the General Agreement on Tariffs and Trade (GATT). Japan had proposed this round in 1971 out of its concern over rising barriers to its exports. Put another way, in many respects Japan has taken the lead in attempts to lower barriers to international trade. In fact, the Tokyo Round did not get underway until early 1975 due to delays in the United States Congress over passing enabling legislation for the American president to begin negotiations. The round was successful in many ways. International tariffs on a wide range of manufactured goods (which Japan was especially strong in) were reduced by a third. There was some agreement with limiting the use of non-tariff barriers to international trade, such as the import quotas on textiles that Japan had submitted to.

Japanese leaders in the Ministry of Finance cooperated in efforts to reduce Japan's trade surplus with the West. Repeatedly, in the late 1970s, they agreed to increase the yen's value still more, making Japanese goods more expensive overseas and provoking complaints from Japanese exporters. Japan also agreed to several domestic spending programmes, in the spirit of Tanaka's original initiative,[6] that were to have led to increased Japanese demand for Western goods. The financial leaders lifted many restrictions on foreign investment in Japan, again in response to Western urging. In early 1977, Japan promised the United States that its economy would grow by at least 7 per cent per year, so that Japanese demand for foreign goods would increase. Not only would this increasing demand reduce Japan's trade surplus, it would also boost production and hence prosperity around the world.

This last promise alarmed many conservatives in Japan, such as Ōhira Masayoshi, who feared that the government would have to increase its spending beyond its means. In a classical exercise of the Yoshida Doctrine, these men compelled the Miki Cabinet to agree that defence expenditures should not exceed 1 per cent of Japan's Gross National

Product.[7] Ōhira, who would become Prime Minister in December 1978, spoke of a 'Comprehensive Security Strategy' (*Sōgō ansen hoshō senryaku*). It emphasized that a nation could be made secure through means other than purely military. Ōhira also meant that Japan could contribute more to the Western alliance by acting as an economic 'locomotive' for the world than building up its armed forces. As if to make the point, Ōhira sponsored a large developmental aid package for the Association of Southeast Asian Nations (ASEAN), such as Indonesia, Thailand, and the Philippines.

Ōhira's strategy was popular with many Japanese, but his timing was unfortunate. In December 1978, the oil-producing countries hiked oil prices by nearly 15 per cent, producing a new oil crisis. Leaders of the industrial nations met at Tokyo in June 1979 to coordinate a response, which would centre around each nation sharply reducing its imports of oil. This was difficult for Japan. Europe had oil reserves of its own in the North Sea, and America was still a substantial producer. Both could reduce their imports of oil without cutting usage very much. Japan, without any oil of its own, was reluctant to commit itself to a fixed percentage of reduced oil imports, again placing it in isolation from the other industrial states. Only American assistance permitted an overall agreement at Tokyo. As well, reduced oil usage meant that Japan might not reach its 7 per cent growth target, and that Japan's trade surplus would not shrink much.

To make matters worse, the overall global situation worsened dramatically by the end of 1979. An Islamic theocracy had come to power in Iran earlier in the year. Its relations with the United States, never good, reached a crisis in November when militants seized the American embassy, took its occupants hostage, and subjected them to public humiliation. When President Jimmy Carter immediately embargoed imports of Iranian oil and called upon America's allies to do the same, Ōhira was placed in a dilemma. Japan not only imported a significant portion of its oil from Iran, it also had invested heavily in the construction of a petrochemical complex at Bandar Shahpur in the wake of its oil difficulties during the 1973 crisis. Yet to defy Carter risked great damage to the alliance with America, where public emotions were running high against Iranians. That American auto-workers were preparing to pulverize Japanese-made automobiles to protest the continued trade surplus hardly helped.

Ōhira understandably tried to temporize, but the Iranians left him little choice. In part to make up for the end of American purchases, they demanded that Japan pay more for their oil than the contract price. Many Japanese oil distributors and consumers thought that Japan should comply. But Ōhira decided to reject the demands, prompting Iran to cut off oil shipments to Japan in April 1980.

By that time Ōhira had a second important decision to make. During the final days of 1979 the Soviet Union had sent large forces into Afghanistan to prop up a faltering pro-Soviet regime. It was the first time that the Red Army had been used outside Europe since early in the Cold War. Coming on the heels of the great disorders in Iran, the move prompted great fear in the West that the USSR was grasping for control of all the Middle East and its invaluable oil.

Ōhira, quite in keeping with Yoshida's earlier ideas, did not believe that the Soviet Union posed any serious threat to Japan, or the West, for that matter. Still, it was important to accommodate the Americans over Afghanistan to preserve a partnership that had benefited Japan immensely for decades. So, once again, Ōhira and Japan hoped for time. His government asked that the Afghanistan question be discussed in the United Nations' Security Council before any hasty steps were taken. But the Soviets hastily vetoed that proposal. Carter called for economic sanctions against the USSR, plus a boycott of the 1980 Olympic Games, which were scheduled to be held in Moscow. Here, Ōhira's choice was easier. Earlier he had played a role in refusing to provide government guarantees to the large project to develop oil fields in Tyumen, Siberia, so there was no sizeable Japanese investment at risk in Soviet territory. Japan's trade with the USSR was insignificant, in part because of the continued impasse over the four islands, the Northern Territories, claimed by both countries. Ōhira agreed to join the sanctions and to send no team to Moscow.

Ōhira had more cause for long-term concern. With the American–Soviet Cold War reborn, there was renewed pressure from Washington for Japan to rearm. By the summer of 1980, the United States had shifted all of its aircraft carriers in the western Pacific to the Indian Ocean. That region of the world showed no signs of settling down, and the death of Josip Tito, leader of Yugoslavia, created fears that southeastern Europe might be tilting toward instability.

These sorts of fears rekindled debate within Japan over what sort of global role it should play in the 1980s. They also contributed to the rising political fortunes of Nakasone Yasuhiro, who would be Prime Minister for much of that decade. Nakasone was a political professional who was much closer in his thinking to the more aggressive nationalism of Hatoyama and Kishi than the Yoshida school. He and his followers bristled at Japan's image of an economic giant and political dwarf. They chafed under Article Nine of the Constitution, which appeared to bar any meaningful military role for Japan under any circumstances. They objected to the arbitrary ceiling of '1 per cent' of Gross National Product that restricted defence spending.

A number of concerns motivated these new hawks. In part, they sincerely felt that Japan owed the West, especially the United States, political and military help. They often heaped scorn upon the Japanese public, which they believed had no conception of the future dangers facing the nation. In part, they feared a new era of global instability. There was a proliferation of novels about a Soviet invasion of Japan, such as Ishihara Shintarō's *National ruin* (*Bōkoku*). Other commentators saw a more likely and even more perilous future. If global conditions deteriorated badly and all nations again turned inward, if the 1980s became the 1930s, the United States might be injured but not endangered. Japan, on the other hand, could see its economic success vanish instantly and itself turned into an economic and political dwarf overnight. As well, though, the hawks hoped that a politically assertive Japan could also more successfully resist the now chronic American pressure over the trade-surplus issue. They remained sensitive to the legacy of the Occupation, and believed that many Americans took Japan for granted. This sensitivity almost inevitably extended to the question of race. It was hard to ignore the fact that of all the members of the industrial nations' club, the OECD, only one – Japan – was not white. Every time that Japan's position was alone, as during the second oil crisis, this issue received great play in Japan.

To a degree, every political leader in the Liberal Democratic Party had these concerns,[8] and that was one reason to agree to Nakasone's premiership. Nakasone, no matter what he actually did inside Japan, was quite useful in presenting a more politically vigorous face outside Japan. In turn, Nakasone was determined to use his global stature to secure reforms within his country and party.

He got off to a quick start. In early 1983 he visited American President Ronald Reagan, bearing gifts. Japan had approved the transfer of military technology to the United States. Publicly, Nakasone declared that Japan would become a real military partner in the Cold War, promising that in times of need Japanese forces would gain control of the Sea of Japan and protect Japanese airspace (and hence a good portion of the Northwest Pacific) from long-range Soviet bombers, referring to Japan as an 'aircraft carrier' that could not be sunk – strong words for the leader of a country with a 'Peace Constitution'. He also approved significant joint exercises between the Self-Defence Forces and the American military, and promised to break through the limit of 1 per cent of Gross National Product on Japanese defence spending.

In short, Nakasone offered to overthrow some of the central premises of the Yoshida Doctrine governing Japan's role in the world. As Japan had achieved modern (or 'high-technology') industrial capability, it had increasingly embraced the Yoshida Doctrine. In 1967, the Satō Government had endorsed a ban on exports of arms to nations which were communist, involved in international disputes, or subject to sanctions by the United Nations. In 1976, these had been expanded to a general ban on all arms exports, including military technology. But the revival of the Cold War and consequent rise of Nakasone reversed rules not even a decade old.[9]

They reversed rules, perhaps, but in the end Nakasone did little to alter Japan's actual role in the world. He would leave office in 1987 with only marginal changes to point to. In part, this was due to the strong sentiment of the Japanese people against change, sentiments shared by the leftist parties and by a large proportion of the LDP. The Left, especially the teachers' unions, bitterly and successfully fought Nakasone's attempts to strengthen a Japanese sense of nationalism through revised school curricula. The arbitrary 1 per cent ceiling on defence spending was breached only once, in 1987, and then only barely and only because Japan's Gross National Product was somewhat smaller than predicted. These were only token reforms. Substantive ones went nowhere. Nakasone failed to end restrictions on the Self-Defence Forces limiting them to 'defensive' weaponry. He could not grant their leader cabinet-level status. He could not have the actual terms of the 1960 Security Treaty even reconsidered, much less the 'no-war' Article of the Constitution. Although Japan ended the 1980s with a much closer working relationship with the American military, this relationship did nothing materially to alter Japan's influence or status within the Western alliance. This was painfully apparent during the West's limited naval intervention after the Iran–Iraq War, which had begun in 1980, escalated to the 'Tanker War' by mid-decade. Japan certainly had an interest in ensuring that oil tankers could sail in the disputed Persian Gulf area, but Nakasone failed to win authorization to send minesweepers there despite assurances that they would not be involved in any belligerent operations.

In part, too, Nakasone's failure to move Japan away from the Yoshida Doctrine was due to international resistance to a wider role for Japan. Here again, Nakasone's ideas of educational reform stirred protest. The Chinese were furious that Japan still admitted no war guilt for the 8 years of rapacious occupation after 1937, fury shared by the inhabitants of Hong Kong and Singapore. The South Koreans remained angered at the way text books in Japanese schools whitewashed the harshness of Japan's colonial occupation from 1910 to 1945.

And, in part, even those who shared Nakasone's criticism of the Yoshida Doctrine often disagreed with his plans for an even closer, and military, association with the United States. Indeed, they believed that a quite opposite approach was in order. Often, this was reflected in books by popular writers, such as Etō Jun's *The American–Japanese War is not over* (*Nichi-Bei sensō wa owatte inai*) which argued that 'war' had become rivalry between high-technology industries organized along national lines.

More famous, because it was translated into English and widely circulated throughout America, was *The Japan that can say 'No'* (*'No' to ieru Nihon*) by Ishihara Shintarō, with parts contributed by Akio Morita of the Sony Corporation. Ishihara painted a grim picture of a Japan that had outgrown its old role as subservient mistress to the United States. It was time for Tokyo to stand up to American pressure in trade, and indeed all matters, to truly end the Occupation era and restore Japan to full sovereignty. In a biting reference to the American oil embargo against Japan five months before the Imperial Navy's attack on Pearl Harbor, Ishihara proclaimed that in the 1980s the American economy depended upon Japanese-made computer microchips to survive even more than Japan had needed American oil fifty years earlier.

Indeed, in the end, economic friction undermined what goodwill Nakasone had won in Washington through his military gestures. Few episodes illustrate the resulting frustration on both sides of the Pacific better than the FSX affair. The FSX (Fighter Support Experimental) was a fighter aircraft intended to replace an aging plane built in Japan through the 1970s. Unlike the old plane, the FSX would be designed not only for supporting ground troops, but would also be used against naval targets, quite in keeping with Nakasone's new role for Japan in the Western alliance.

The Americans were pleased with the role, but preferred that Japan buy an American-made aircraft and modify it. In this way, Japan could reduce its still growing trade surplus with the United States. Modifications proved unfeasible, but in late 1988 both governments signed an accord to co-produce the FSX. American companies would receive nearly half the production work. Japan would bear the entire expense of the plane's development costs and would make all resulting technology available to Washington.

The accord appeared to be an ideal solution for the alliance, but not to many members of the American Congress, the Commerce Department, and labour unions. They argued that the accord was a Trojan horse that would allow Japan access to production techniques and technology in the aerospace industry, virtually the only industry in which the United States still enjoyed a trade surplus with Tokyo. Once Japan learned how to build the FSX, it would soon be competing with American companies (and their workers) for a share of the global aerospace market. These critics, pointing to past Japanese successes in automobiles, electronics, and semiconductors, found a ready audience among many Americans. Congress passed legislation greatly restricting transfer of aerospace technology to Japan and requiring firming guarantees of American co-production work. This unusually direct intervention into United States–Japan relations provoked President George Bush to veto the legislation in July 1989.

The FSX affair only symbolized how acrimonious those relations had become by the end of the 1980s. At the centre of that acrimony was the still-ballooning Japanese trade surplus: over US$50 billion a year with the United States alone. During the FSX debates, one member of Congress argued that there had to have been a conspiracy of 'shogun-like characters'[10] responsible for the destruction of American industry after industry and the resulting deficit. The truth was even bleaker. For nearly all of the 1980s, the greatest diplomatic activity within the Japanese–American alliance, and so in Japanese foreign relations, had deliberately focused on reducing Tokyo's trade surplus. And all that diplomacy had failed.

The trade-surplus problem was hardly new. Already in the late 1970s Japan had promised to reduce it by increasing its domestic economic growth and hence demand for imports. This strategy was logical, but it ran foul of the second oil crisis of 1979. Japan did increase its imports, but of oil from the Middle East instead of products from the United States.

That crisis also worsened inflation throughout the world. Japan immediately adopted measures to combat it by raising interest rates and controlling domestic spending. These measures swiftly brought inflation under control, but created new difficulties with America. Japan's discipline had done nothing to increase American imports. And because Tokyo had brought inflation under control more successfully than Washington, the prices of Japanese goods remained low while America's inflation drove up the price of American products. As a result, the trade surplus actually grew in the early 1980s, reaching an unprecedented amount of US$44.2 billion in 1984.

This unfortunate situation worsened with the enactment of President Ronald Reagan's economic programmes in the United States. The centrepiece was a large reduction of American income taxes and its effects were twofold. First, the ensuing boom in spending led to a large surge of imports from Japan, bloating the trade surplus further. Second, since Washington was unable to reduce its spending (though it had no difficulty slashing its tax revenues), it ran huge budget deficits. Increasingly through the 1980s, Japan funded America's spendthrift ways, with the colossal amount of US$103 billion flowing to the United States in 1985 alone.[11]

The straightforward way to address these problems would have been for Washington to raise taxes and reduce government spending, but this was politically impossible. The other option was to once again seek to raise the value of the yen and lower the dollar so that Japanese goods would be more expensive and American ones cheaper, hence more competitive. For several years Reagan resisted this option, viscerally believing that a 'strong' dollar (high in value) was an important symbol of a strong America in general. In the meantime, Japanese imports scored large gains in the important automotive, machinery, and electronics markets of the United States, increasing demands for protection against imports of those products.

These pressures came to a head in September 1985. Reagan, despite his vocal commitment to free trade, invoked Section 301 of the Trade Act to impose strict penalties against Japanese imports. This section permitted penalties if the target nation was found to have restricted imports of American goods, in this case tobacco products. In response, Prime Minister Nakasone called upon every Japanese to buy US$100 worth of imports. He visited a department store and ostentatiously bought several neckties (which some sullen Americans pointed out had been made in France). The results of this tokenism were negligible.

So also was the Plaza Agreement[12] that same month. Under it, finance ministers from the industrial powers cooperated to allow the dollar to fall in value while the yen would rise, and by an appreciable amount (over 10 per cent). As a result, Japan's trade surplus was to be reduced. Actually, only Nakasone's popularity was reduced as Japanese exporters and LDP leaders who coveted the prime ministership bitterly criticized another surrender to the Americans. Nakasone deftly met this challenge by appointing one of those leaders, Miyazawa Kiichi, as Finance Minister in July 1986. Miyazawa returned to the old solution of stimulating demand within Japan through government spending, but as Japan entered the 1990s there was no sign of its trade surplus declining. Indeed, in 1988 that surplus had mushroomed to US$95 billion.

The Plaza Agreement did have an effect, however, on Japanese investment overseas, especially in the United States.[13] A stronger yen and weaker dollar meant that American goods indeed were cheaper, but this situation also meant that American land and companies were cheaper, too. The late 1980s saw a series of highly-publicized Japanese acquisitions of American properties, from Rockefeller Center in New York to Columbia Pictures of Hollywood to much real estate in Hawaii, prompting dark American humour about how much of Pearl Harbor was left to the United States.

In other ways Pearl Harbor also symbolized continued difficulties in Japan's foreign relations. From there, the United States still projected power globally. In contrast, Japan continued to be a political dwarf during the Persian Gulf War of 1990–1 and the concomitant end of the Cold War.

When Iraq invaded Kuwait in August 1990, Japan's initial reaction was swift. Unlike the dithering of the oil crisis of 1973, this time Tokyo agreed to join an international agreement to not import oil from Iraq or its occupied victim and to refuse additional investment there. Japan remained reluctant to do more. Prime Minister Kaifu Toshiki refused to provide a specific amount of medical supplies and relief for Kuwaiti refugees for nearly a month. Joining in a multinational military force organized by the United States to defend Saudi Arabia, Kuwait's neighbour, was out of the question.

This aloof stance was instantly ridiculed by many in the West as craven 'chequebook diplomacy'. Some asked why American soldiers should be asked to fight and die to free Kuwait to sell more oil – to Japan. Even scholars friendly to Tokyo wondered how Japan could justify its aspirations for a seat on the United Nations' Security Council when it so pointedly refused to accept a role in international peacekeeping efforts. These concerns only grew as it became clear that Japan would have difficulty delivering medicine and relief supplies. The Kaifu Cabinet refused to use the Self-Defence Forces to deliver these, preferring to charter private aircraft and shipping. But no planes were chartered, and maritime unions refused to load ships bound for the war zone. A call for Japanese doctors to go to Saudi Arabia went virtually unanswered.

Stung, Kaifu proposed the 'United Nations Peace Cooperation Law' to the Diet in mid-October. This proposal would have created a 'peace cooperation corps' under the Prime Minister, controlled by civilians, that could draw on SDF units to be used on strictly non-military operations. Weak as it was, the bill drew fire from the Foreign Ministry, which objected to any use of SDF units outside Japan, and the majority of Diet members, who opposed the entire concept. Even Kaifu's own Liberal Democratic Party disliked the bill, except for Ozawa Ichirō, who had emerged as spokesman for the hawkish wing of the Liberal Democratic Party, favouring a more aggressive role for Japan in the world and the constitutional revision such a role would require. The bill was withdrawn, and out came the chequebook. By the war's conclusion in February 1991, Japan had sent nearly US$13 billion in support. Even so, the Diet had stipulated the money was not to be used for arms or ammunition. It is true that Japan sent minesweepers into the Persian Gulf in April. But this was well after the shooting had stopped. Even then, the ships were sent without a support vessel transporting helicopters, since these were deemed too 'militant'.

The widespread international dissatisfaction with Japan's reluctant and limited response did lead to one small step away from the Yoshida Doctrine and toward a real sense of comprehensive security. The 'International Peace Cooperation Law' of 1992 permitted units of the Self-Defence Forces (and private volunteers) to participate in United Nations peacekeeping operations. To ensure passage, however, the law was even more restrictive than Kaifu's earlier bill. The SDF could not join in such operations with any military component, for example, monitoring ceasefires or even confiscating abandoned weapons. If any operation should threaten to involve military operations, Tokyo would have the right to withdraw from the UN force unilaterally. In addition, the operations literally had to be UN affairs. The Persian Gulf War was not. Japanese SDF units departed for Cambodia in late 1992 to join a UN force there. Newsworthy as that was – the first use of Japanese troops in Southeast Asia since 1945 – the law also remains a demonstration of the formidable domestic obstacles to any meaningful role for Japan in global political and military affairs.

Much the same can be said of Japan's glacier-like response to the end of the Cold War and breakup of the Soviet Union. Tokyo remained the slowest of Western capitals to acknowledge that Mikhail Gorbachev had brought fundamental change to the USSR after coming to power in 1985. Far from assisting Gorbachev's reform efforts, Japanese leaders insisted upon a resolution to the Northern Territories dispute before serious discussions on wider issues could begin.

This stance guaranteed impasse. Gorbachev, deeply concerned about secessionist tendencies in the Baltic states and the possible breakup of the Soviet Union, could hardly afford to yield any territory to Japan, even the four rather miserable islands of the Northern Territories.[14] Yet, for Japan, those islands had acquired a significance far beyond their actual value. Huge numbers of Japanese signed petitions demanding their return. Many intellectuals spoke of the Soviet deadlock as the Asian equivalent of the Berlin Wall.

The reason for this depth of feeling is easy to see. For many Japanese, the Cold War could not be over until the islands were returned. For many Japanese, the Cold War symbolized a long period of incomplete sovereignty for Japan, sovereignty that literally cannot be restored until the Northern Territories are. Bitterness also remains in the mid-1990s because the Soviets captured over 550,000 Japanese soldiers at the end of the Pacific War and used them as labourers. Possibly as many as 70,000 died and Japan wants an apology and accounting. To no one's great surprise, the territories continued to block any development in Japanese–Russian relations, forcing Russian President Boris Yeltsin to cancel a visit to Tokyo in late 1992. Japan's reluctance to compromise particularly irritated the Europeans, who feared floods of refugees and renewed violence if the Russian economy collapsed for want of sufficient Western (including Japanese) assistance.[15]

The hand of the past still lays heavily on Japan's recent relations with its Asian neighbours, too. After protracted delays, the Japanese Government finally admitted, in 1993, that it had actively sponsored the 'recruitment' of Korean and other Asian 'comfort women' to serve personnel of the Imperial Army during the Pacific War. Even by then, however, there was no official apology to China for that war's occupation, to Beijing's obvious disappointment. While Southeast Asian nations welcome Japanese efforts to stabilize the chaos in Cambodia, they remain wary of both Japan's reliability in peacekeeping operations and, paradoxically, its responsibility in the use of military power if Japan ever again gains that power. It is a paradox also found in Washington, which wants greater Japanese power and responsibility but is unsure Japan can be trusted with either. Even in 1994, nearly 50,000 American soldiers and sailors remained based in Japan.

Perhaps more troubling are the Koreans who remain in Japan. Holdovers from the colonial era, plus refugees from the time of the Korean War, these Koreans had not been extended Japanese citizenship rights. Only by 1990 did the Japanese Government move to establish a legal basis for ethnic Koreans living in Japan, and then only for the third generation of these Koreans. Only by 1993 did it stop fingerprinting these residents, ironically, just as the Americans became concerned that some Korean residents in Japan were funnelling large supplies of cash to the isolated North Korean (communist) regime, which was considering building its own atomic bomb, an issue that blossomed into crisis by early 1994.

More broadly, Japan has tightened its immigration laws. Many labourers from Asia and the Middle East were finding their way to prosperous Japan in search of jobs. Many Japanese pointed to the West German experience with 'guest workers' who became permanent residents and a permanent source of social friction. There is little desire to follow Germany's example. By the same token, Japanese scholars and journalists have been following the discussions of 'multiculturalism' in America with growing interest and alarm. They are drawn to the idea of a well-defined cultural identity, as championed by American opponents of immigration. But these same Americans define a cultural identity that is unmistakably Western and white, with unsettling implications for the future of a Japan that is somehow in the West but not of it.

In many respects, the Yoshida Doctrine is as alive as ever. There appears to be no end to Japan's trade surpluses with the rest of the West, nor to Japan's fundamental reluctance to play any international role except successful merchant. Many Japanese are fundamentally averse to revising the 'no-war' clause of the Constitution. They tend to favour a greater international role through the United Nations, including a permanent seat for Japan on the UN Security Council. It is not clear how such a position could be squared with the UN Charter's dictates that member nations make available their military forces 'for the purpose of maintaining international peace and security'.[16]

Increasingly as the twentieth century draws to an end, Japanese scholars are asking how long the Yoshida Doctrine and the international system that allows it can continue. Some point to a global trading system that sees Japan exporting goods to Europe and America, which sell arms to the oil-producing countries, which market oil to Japan. Although a considerable oversimplification, this triangle of trade has unsettling implications for Japan. Indeed, another popular image of the Japan of the twenty-first century is equally unsettling. Unlike Great Britain of the nineteenth century or America of the twentieth, Japan has no military power, nor political institutions capable of employing military power, nor the domestic wherewithal or will to create a basis for a self-sufficient Japanese military. Unlike Britain or America, Japan has no culture or language easily transposed upon others worldwide. Indeed one of the great absences of the history of modern Japan's foreign relations has been the nearly complete lack of the spread of a global understanding of Japanese culture. Unlike Britain or America, Japan has no ideology comparable to free trade or human rights or, for that matter, the idea of a 'white man's burden' to spread that ideology and sense of civilization to serve as a catalyst for the exercise of global influence. Much more persuasive is an analogy suggested by diplomat Okazaki Hisahiko: Japan as the equivalent of the seventeenth-century Dutch: hugely prosperous, but vulnerable to protectionism and, for that matter, 'a sound thrashing', anytime a rival desires to administer one. Okazaki points to an England that aided Holland through its long struggles against the Catholic powers of Europe, at great expense to England in money and blood while the Dutch profited handsomely. Yet within five years of securing peace in that wider, longer conflict, the English had turned on the Dutch and ruined them. It is a metaphor to cause nightmares in Tokyo.

Rather more plausibly, however, the Japanese can imagine a world after the Cold War in which order has broken down outside the confines of the industrialized nations and the rules of international commerce no longer apply. They can likewise imagine a new emergence of trading blocs within those confines, as the European Community pursues greater integration and the Americans build on initiatives such as the North American Free Trade Agreement (NAFTA). Japan, which has gained immensely from the creation of a global system of free trade, would lose immensely in any return to regional mercantilism. Few Japanese relish the prospect of a return to Tokugawa-style isolation, even one that would see Japan as the economic leader of an East Asian bloc. They understand that there is no recourse to isolation today any more than there ever was in 1853. Japan remains destined to play a pivotal role in the history of humankind. To better understand that role from Perry to present has been this study's purpose.

Notes

[1] In addition, the United States actively encouraged the transfer of technology from American to Japanese companies. Nearly three-quarters of Japan's licences for such transfers came from American sources.

[2] In this case, Maurice Stans, who became Nixon's Secretary of Commerce. Other well-known business leaders who would repeat this tactic included John Connally and, later, Lee Iococca.

[3] Destler, I.M. 1979: *Conflict in Japanese–American relations, 1969–1971.* Ithaca, NY: Cornell University Press, 180.

[4] Japanese development of Siberia, which would strengthen the USSR economically, was, after all, hardly compatible with the anti-hegemony pledge Japan had made with China.

[5] From 308 yen to the dollar to 277, a change of about 10 per cent. Before the second Nixon shock of August 1971, 360 yen equalled 1 dollar.

[6] Tanaka himself had been forced out of office in late 1974 due to the Lockheed scandal involving improper use of funds in aircraft purchases, though he remained a 'kingmaker' in the LDP well afterwards.

[7] Gross National Product, or GNP, is the total value of all goods and services produced in a nation.

[8] Some, closer to the Yoshida school, continued to believe that the Americans radically overestimated the Soviet threat, though they were willing to see Nakasone propitiate the new Cold Warriors in Washington in the wider interests of the alliance so long as the fundamentals of the Yoshida Doctrine went unaltered.

[9] Satō also had enunciated the 'Three Non-Nuclear Principles' which barred Tokyo from possessing, producing, or introducing nuclear weapons into Japan. This, too, ran foul of Cold War developments, although different ones. Most Japanese political leaders were reluctant to ratify the Nuclear Non-Proliferation Treaty after Japan signed it in early 1970. The reason had little to do with Japan's military, however. The first oil crisis of 1973 convinced even the Japanese public that Japan had to find alternative sources of energy, and nuclear reactors seemed quite promising. Japan had the technology to build and operate the reactors. But an old theme – lack of resources – continued to haunt the programme. Japan had no reactor-grade 'enriched' uranium, relying upon the United States for its supply. There was a solution to this problem: 'breeder' reactors which actually produced more nuclear fuel than they used. But that fuel could also be made into nuclear weapons. The Japanese Left challenged the breeder reactor idea, claiming that it implicitly violated the 'no-war' article of Japan's constitution. The conservatives were not going to allow the article to block their hopes for nuclear

energy, and had announced that Japan would reserve the right to develop that energy for peaceful purposes. This stance was fine in principle, but it complicated Japan's acceptance of the Non-Proliferation Treaty, because the treaty insisted that signatories yield the right of inspection to international teams. To the Japanese, this smacked of extraterritoriality, and indeed of an 'unequal treaty', since the European powers had obtained a waiver from granting such a right. Even so, Japan finally ratified the treaty in 1976.

[10] Representative Dana Rohrabacher, quoted in *Wall Street Journal*, 16 May 1989, A9.

[11] To give a sense of proportion: in 1981 the United States had invested $141 billion more overseas than non-Americans had invested in the United States. By 1985, Americans had invested $111 billion less, a startling reversal in both size and speed.

[12] The agreement was named after the Plaza Hotel, where it was negotiated.

[13] In addition, Japan moved past Great Britain and France to become, along with West Germany, the second largest holder of voting rights in the International Monetary Fund.

[14] Actually, one of the four is a tiny group of islands, the Habomais.

[15] Yeltsin finally visited Japan in October 1993. He extended assurances that Russian forces would leave the disputed islands, but no progress was made on resolving the territorial issue.

[16] Article 43, Charter of the United Nations.

Suggestions for additional reading

There are few surveys of Japan's foreign relations, at least in English. Those in existence include William G. Beasley, *Japanese imperialism, 1868–1945* (Oxford: Clarendon Press, 1987); Ian H. Nish, *Japanese foreign policy, 1869–1942: Kasumigaseki to Miyakezake* (London: R. & K. Paul, 1977); Kajima Morinosuke, *The diplomacy of Japan, 1894–1922*, 3 vols (Tokyo: Kajima Institute of International Peace, 1966–8); John Hunter Boyle, *Modern Japan: The American nexus* (Forth Worth, TX: Harcourt Brace Jovanovich, 1993) with a focus on United States–Japan relations; and Sydney Giffard, *Japan among the powers, 1890–1990* (New Haven: Yale University Press, 1994).

Specialized studies of those relations are somewhat more plentiful, though not for the early period during, and shortly after, the Meiji Restoration. A good place to begin is with E.H. Norman's works, introduced by John W. Dower, *Origins of the modern Japanese state* (New York: Pantheon, 1975); William G. Beasley, *The Meiji Restoration* (Stanford: Stanford University Press, 1972); chapters by Beasley and Akira Iriye in volume 5 of *The Cambridge history of Japan* (Cambridge: Cambridge University Press, 1989); and Marius Jansen's essay in Robert E. Ward (ed.), *Political development in modern Japan* (Princeton: Princeton University Press, 1968).

Japan's period of high imperialism, to nearly the start of the First World War, can be investigated in Hilary Conroy, *The Japanese seizure of Korea, 1868–1910* (Philadelphia: University of Pennsylvania Press, 1960); Ian Nish, *The Anglo-Japanese Alliance: The diplomacy of two island empires, 1894–1907* (London: Athlone Press, 1985) and his *The origins of the Russo-Japanese War* (New York: Longman, 1985); Roger F. Hackett, *Yamagata Aritomo in the rise of modern Japan, 1838–1922* (Cambridge: Harvard University Press, 1971); Tatsuji Takeuchi, *War and diplomacy in the Japanese Empire* (Garden City, NY: Doubleday, 1935); Bonnie Oh's chapter in Akira Iriye (ed.), *The Chinese and the Japanese* (Princeton: Princeton University Press, 1980); and the Asian chapters of William L. Langer, *Diplomacy of imperialism, 1890–1902* (New York: Knopf, 1951).

The beginnings of difficulties with America and the complications of the war and postwar adjustments can be traced in Akira Iriye, *Pacific estrangement* (Cambridge: Harvard University Press, 1972); Tetsuo Najita, *Hara Kei and the politics of compromise, 1905–1915* (Cambridge: Harvard University Press, 1967); James W. Morley, *The Japanese thrust into Siberia, 1918* (New York: Columbia University Press, 1957); Ian Nish, *Alliance in decline: A study in Anglo-Japanese relations, 1908–23* (London: Athlone Press, 1972); Roger Dingman, *Power in the Pacific: The origins of naval arms limitation, 1914–1922* (Chicago: University of Chicago Press, 1976); Akira Iriye, *After imperialism: The search for a new order in the Far East, 1921–1931* (Cambridge: Harvard University Press, 1965); William F. Morton, *Tanaka Gi'ichi and Japan's China policy* (Folkestone, England: Dawson, 1980); Bamba Nobuya, *Japanese diplomacy in a dilemma: New light on Japan's China policy, 1924–1929* (Vancouver: University of British Columbia Press, 1972); various works of Sadao Asada, such as 'From Washington to London: The Imperial Japanese Navy and the Politics of Naval Limitation, 1921–30,' *Diplomacy & Statecraft* (Nov. 1993): 147–91; Masaru Ikei's study of Ugaki Kazushige in Akira Iriye (ed.), *The Chinese and the Japanese* (Princeton: Princeton University Press, 1980); and chapters by Kato and Iriye in Bernard Silberman (ed.), *Japan in crisis: Essays on Taisho democracy* (Princeton: Princeton University Press, 1974).

The many roads to war of the 1930s have understandably received a great deal of attention from historians. One indispensable source, of special value because of its Japanese scholarship, is the multi-volume collection translated, under the editorship of James W. Morley and Columbia University Press, New York, from the *Taiheiyō sensō e no michi* studies. These volumes are: *Japan erupts: The London Naval Conference and the Manchurian incident, 1928–1932* (1984); *The China quagmire: Japan's expansion on the Asian continent, 1933–1941* (1983); *Deterrent diplomacy: Japan, Germany, and the USSR, 1935–1940* (1976); and *The fateful choice: Japan's advance into Southeast Asia, 1939–1941* (1980). Other good studies for this period include Dorothy Borg and Shumpei Okamoto with Dale Finlayson (eds.), *Pearl Harbor as history: Japanese–American relations, 1931–1941* (New York: Columbia University Press, 1973); James B. Crowley, *Japan's quest for autonomy: National security and foreign policy, 1930–1938* (Princeton: Princeton University Press, 1966); Stephen E. Pelz, *Race to Pearl Harbor: The failure of the Second London Naval Conference and the onset of World War Two* (Cambridge: Harvard University Press, 1974); Akira Iriye, *The Origins of the Second World War* (New York: Longman, 1987); chapters by Ikuhiko Hata and Alvin Coox in volume 6 of *The Cambridge history of Japan: The Twentieth Century* (Cambridge: Cambridge University Press, 1988); and this author's *Japan prepares for total war: The search for economic security, 1919–1941* (Ithaca: Cornell University Press, 1987).

The war years themselves, curiously, have not received ample attention from academic historians. But the few works completed are insightful. These are: Akira Iriye, *Power and culture: The Japanese–American War, 1941–1945* (Cambridge: Harvard University Press, 1981); Leon V. Sigal, *Fighting to a finish: The politics of war termination in the United States and Japan, 1945* (Ithaca: Cornell University Press, 1988); and Robert J.C. Butow, *Japan's decision to surrender* (Stanford: Stanford University Press, 1954).

Given the paucity of documents released by the Japanese Government for and during the long postwar era, true historical studies of this period are somewhat tentative. But there are a number of extraordinarily thoughtful ones written under such constraints, such as: John W. Dower, *Empire and aftermath: Yoshida Shigeru and the Japanese experience, 1878–1954* (Cambridge: Harvard University Press, 1979), a study that spans prewar, wartime, and the Occupation years, as does, from an economic policy perspective, Chalmers Johnson, *MITI and the Japanese miracle: The growth of industrial policy, 1925–1975* (Stanford: Stanford University Press, 1982). Howard B. Schonberger, *Aftermath of war: Americans and the remaking of Japan, 1945–1952* (Kent, OH: Kent State, 1989) and Richard B. Finn, *Winners in peace: MacArthur, Yoshida and postwar Japan* (Berkeley: University of California Press, 1992) are especially good for the Occupation period. Additional concerns are covered in Marc S. Gallicchio, *The Cold War begins in Asia: American East Asian policy and the fall of the Japanese Empire* (New York: Columbia University Press, 1988); Donald C. Hellmann, *Japanese foreign policy and domestic politics: The peace agreement with the Soviet Union* (Berkeley: University of California Press, 1969); I.M. Destler, *The textile wrangle: Conflict in Japanese–American relations, 1969–1971* (Ithaca: Cornell University Press, 1979); Martin E. Weinstein, *Japan's postwar defense policy, 1947–1968* (New York: Columbia University Press, 1971); and George R. Packard II, *Protest in Tokyo* (Princeton: Princeton University Press, 1966). More contemporary coverage can be found in Reinhard Drifte, *Japan's foreign policy* (London: Royal Institute of International Affairs, 1990); F.C. Langdon, *Japan's foreign policy* (Vancouver: University of British Columbia Press, 1973), and, for financial matters, the collaborative effort by Paul Volcker and Toyō Gyōten, *Changing fortunes: The world's money and the threat to American leadership* (New York: Times Books, 1992). Three excellent surveys of the entire postwar era are Kenneth Pyle, *The Japanese question: power and purpose in a new era* (Washington: American Enterprise Institute, 1992); Roger Buckley, *US–Japan Alliance diplomacy, 1945–1990* (Cambridge: Cambridge University Press, 1992); and John Welfield, *An empire in eclipse: Japan in the postwar American alliance system* (London: Athlone Press, 1988).

Index